ANATOMY IN OUTLINE

George J. Furman, MD, MS, FACS

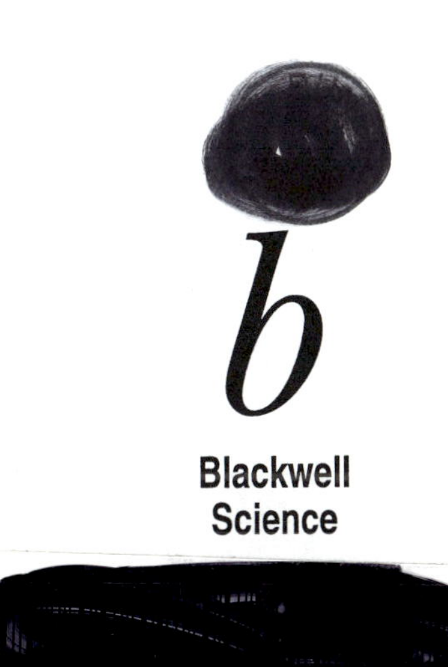
Blackwell Science

EDITORIAL OFFICES:

Commerce Place, 350 Main Street, Malden, Massachusetts 02148, USA
Osney Mead, Oxford OX2 0EL, England
25 John Street, London WC1N 2BL, England
23 Ainslie Place, Edinburgh EH3 6AJ, Scotland
54 University Street, Carlton, Victoria 3053, Australia
Other Editorial Offices:
Blackwell Wissenschafts-Verlag GmbH, Kurfürstendamm 57, 10707 Berlin, Germany
Blackwell Science KK, MG Kodenmacho Building, 7-10 Kodenmacho Nihombashi, Chuo-ku, Tokyo 104, Japan

DISTRIBUTORS:

USA Blackwell Science, Inc.
Commerce Place
350 Main Street
Malden, Massachusetts 02148
(Telephone orders: 800-215-1000 or 781-388-8250; fax orders: 781-388-8270)

Canada Login Brothers Book Company
324 Saulteaux Crescent
Winnipeg, Manitoba R3J 3T2
(Telephone orders: 204-837-2987)

Australia Blackwell Science Pty, Ltd.
54 University Street
Carlton, Victoria 3053
(Telephone orders: 03-9347-0300;
fax orders: 03-9349-3016)

Outside North America and Australia
Blackwell Science, Ltd.
c/o Marston Book Services, Ltd.
P.O. Box 269
Abingdon
Oxon OX14 4YN
England
(Telephone orders: 44-01235-465500;
fax orders: 44-01235-465555)

Associate Editor: Laura Berendson
Production: Irene Herlihy
Manufacturing: Lisa Flanagan
Marketing Manager: Toni Fournier

Cover design by Electronic Illustrators Group
Typeset by Modern Graphics, Inc.
Printed and bound by Edwards Brothers/Ann Arbor

Printed in the United States of America
00 01 02 03 5 4 3 2 1

Library of Congress Cataloging-in-Publication Data

Furman, George J., 1930-
Anatomy in outline / by George J. Furman.
p. ; cm.
Includes citations to color plates in Atlas of human anatomy, 2nd ed., by Frank H. Netter.
Includes index.
ISBN 0-86542-592-2
1. Human anatomy--Outlines, syllabi, etc. I. Netter, Frank H. (Frank Henry), 1906-
Atlas of human anatomy. II. Title.
[DNLM: 1. Anatomy--Outlines. QS 18.2 F986a 2001]
QM31 .F95 2001
611--dc21

00-033674

Preface

Anatomy is the essential framework for the medical sciences. *Anatomy in Outline* organizes basic anatomical information to aid medical students in study, review, surgical rotations, and preparation for board examinations. It is not intended as a substitute for standard textbooks, atlases, or laboratory dissection.

In the *Outline* the author, an anatomist and surgeon, brings together fundamental anatomical descriptions along with selected clinical correlations in a systematic, concise, and readily retrievable outline format. Clinical correlations are printed in green type for quick reference.

The *Outline* employs standard nomenclature in addition to the terms commonly employed by surgeons. It includes occasional practical references to related fields such as embryology, histology, and the neurosciences. Parenthetical statements and tables have been added to help bridge the gap between basic gross anatomy and clinical anatomy.

The *Outline* has been formatted for rapid access to information and to facilitate learning:

- Citations to color plates in *Atlas of Human Anatomy*, second edition, by Frank H. Netter are included in the body of the text as well as in the chapter tables of contents. Plate numbers are printed in green type and marked by an arrow icon in the margin.
- Review questions and detailed answers are found at the back of the book.
- Simple line illustrations serving as reminders of basic concepts also appear in the margins, next to their corresponding text.
- Text has been boldfaced and italicized to draw attention to key concepts and to assist the reader in quickly navigating the outline.
- The book utilizes a special binding to lay flat.

Anatomy in Outline, then, is intended to be the best organizational guide for the study and review of anatomy—the framework of medicine.

George J. Furman, MD, MS, FACS

Acknowledgments

I am indebted to my teachers and students for their inspiration and encouragement over the years. I wish to thank those who provided suggestions, including Joanne Keenan and Bob Clift. I am particularly grateful to Natalie Holt of Yale University for contributing the Question & Answer section and to my wife, Nancy, for her patience.

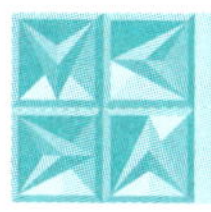

Table of Contents

General

Chapter 1

Page numbers in black. Netter plate numbers in green.

Planes

I. PLANES

A. Vertical

1. *Anterior to posterior*

i. median or median sagittal—a vertical plane thru the center of the body

ii. sagittal—a plane at **or** parallel to the median plane of the body (i.e., from front to back)

iii. paramedian—a vertical plane passing thru the body (front to back) and lateral to the median sagittal plane (i.e., on the side of the median plane)

B. Side to Side

1. *Coronal or Frontal—a plane at right angles to a median plane (i.e., perpendicular to the median plane of the body). It divides the anatomical part into anterior and posterior portions.*

C. Horizontal

1. *Transverse—a plane at right angles to a coronal plane. It divides the anatomical part into upper and lower portions.*
2. *Transaxial—a plane cut perpendicular to the axial line of the region being described*

➢ II. PLANES RELATED TO THE ABDOMINAL WALL (251)

A. Horizontal

1. *Subcostal—at the inferior margin of the 10th costal cartilage (L3 level)*
2. *Transpyloric—halfway between the jugular notch and the pubic symphysis (L1 level)*
3. *Transumbilical (L3–L4 level)*
4. *Transtubercular (Intertubercular)—between the iliac tubercles (L5 level)*

B. Vertical

1. *Midclavicular (Midinguinal)—from the middle of the clavicle to the middle of the inguinal ligament*
2. *Midsternal—in the middle of the sternum*
3. *Lateral sternal—along the lateral aspects of the sternum*

Aortic Arch

- **I. BRACHIOCEPHALIC (131)—a branch of the aortic arch. The brachiocephalic artery is about 2 inches long and passes posterior to the right sternoclavicular joint.**
  - A. Right Common Carotid (29)—begins behind the sternoclavicular joint and is located in the neck. At the upper border of the thyroid cartilage, it divides into:
    1. *External carotid—supplies the head, face, and neck*
    2. *Internal carotid—supplies the cranial and orbital cavities*
  - B. Right Subclavian
    1. *Thyrocervical trunk* (28, 131)
       - i. inferior thyroid—to the inferior thyroid and parathyroid
         - a. ascending cervical
       - ii. suprascapular—to the muscles around the scapula
       - iii. transverse cervical—to posterior triangle neck muscles
    2. *Vertebral (14, 95, 130, 136, 157)—ascends thru the transverse foramina of vertebrae to enter the foramen magnum—supplies brain stem, spinal cord, and cerebellum*
       - i. basilar—formed by the right and left vertebral arteries and divides into the superior cerebellar and posterior cerebellar on both right and left sides
       - ii. posterior inferior cerebellar—to the cerebellum, choroid plexus, and 4th ventricle
       - iii. spinal arteries—Posterior spinal arteries (right and left) and an anterior spinal artery supply the spinal cord.
       - iv. posterior meningeal—to dura and bone in the posterior fossa
    3. *Internal thoracic (Internal mammary) (168, 176)—passes inferiorly next to the sternum*
    4. *Costocervical trunk*
       - i. superior intercostal
       - ii. deep cervical—to neck muscles
    5. *Axillary artery—a continuation of the subclavian*
- **II. LEFT COMMON CAROTID (131, 182)—begins as a branch of the aortic arch**
- **III. LEFT SUBCLAVIAN (182)—a branch of the aortic arch. It passes posterior to the left sternoclavicular joint.**

Vertebral Arteries

I. COURSE (14)—The vertebral arteries begin at the subclavian arteries, ascend in the foramina of the transverse processes of the cervical vertebrae, and then go between the atlas and the occipital bone to enter the foramen magnum. They emerge in the transverse process of the atlas, run posteromedially, and enter the vertebral canal. They pierce the dura in the cranial cavity. At the junction of the medulla and pons, the vertebral arteries merge to form the basilar artery. The basilar artery then supplies the pons, medulla, mesencephalon, cerebellum, occipital lobe, and medial aspect of the temporal lobe.

II. BRANCHES OF THE VERTEBRAL (132)

A. Spinal—supplies the spinal cord—has anterior and posterior branches to ventral and dorsal horns, respectively

B. Posterior Inferior Cerebellar—supplies the inferior cerebellum, choroid plexus, and lateral medulla

C. Basilar—formed by the right and left vertebrals

III. BRANCHES OF THE BASILAR (133)

A. Anterior Inferior Cerebellar—to the anterior cerebellum and brain stem (**anastomoses** with the posterior)

B. Internal Labyrinthine (Auditory)—to part of the inner ear

C. Pontine—to pons and cerebellum

D. Superior Cerebellar—to pons and cerebellum

E. Posterior Cerebral—to parts of the cortex (temporal and occipital lobes), thalamus, and hypothalamus

IV. CIRCLE OF WILLIS (133)—surrounds the area containing the optic chiasma, pituitary, and mammillary body—composed of branches of the basilar and internal carotid arteries

A. Posterior Cerebral—a continuation of the basilar beyond its junction with the superior cerebellar

B. Posterior Communicating—supplies the optic tract, hypothalamus, and thalamus—connects the right and left posterior cerebrals (of the basilar) and the internal carotid

C. Internal Carotid—at the junction of the anterior cerebral and posterior communicating arteries

D. Anterior Cerebral—a terminal branch of the internal carotid—supplies the optic chiasma, anterior hypothalamus, and part of the frontal and parietal lobes, and joins the anterior communicating

E. Anterior Communicating—connects the anterior cerebrals

Lymphatic System

I. JUGULAR TRUNK (66)—On the right it may empty into the right lymphatic duct. On the left it may empty into the thoracic duct. The trunks may also drain directly into the veins without entering the thoracic or right lymphatic duct.

A. Deep Cervical Nodes (15–30)—located along the carotid artery and internal jugular vein and divided into superior and inferior (supraclavicular) groups. **The deep nodes receive the superficial cervical nodes** and have direct drainage from mucous membranes and respiratory passages located in the head and neck.

1. *Superior group—related to the internal jugular vein. The group is located above the omohyoid muscle.*

i. submandibular—related to the internal jugular vein and inferior mandible—drain the face, lips, and chin

a. submental—related to the internal jugular vein and the mylohyoid muscle—drain the lower lip and oral cavity floor

b. facial—located below the orbit, on the buccinator muscle and on the outer surface of the mandible—drain the eyelids, cheek, nasal skin, and mucosa

ii. parotid—drain the lateral aspects of the face and the parotid gland

a. anterior auricular

2. *Inferior (Supraclavicular) group—located near the subclavian vein and inferior to the omohyoid muscle—receives some lymphatics from the upper limb (apical axillary nodes) and the thoracic wall in addition to those from the head and neck. The inferior group drains the thyroid, trachea, larynx, pharynx, tonsils, upper esophagus, and an upper portion of the breast.*

B. Superficial Cervical Nodes—receive efferents from the occipital, mastoid, and parotid areas and from skin and muscles. The superficial nodes receive efferents from the posterior triangle.

II. RIGHT LYMPHATIC DUCT (197)—drains the right upper limb and the right side of the head, neck, and thorax. It enters the right internal jugular vein or the right subclavian vein.

A. Right Jugular Trunk—drains the right side of the head and neck

B. Right Bronchomediastinal Nodes

1. *Right internal thoracic (mammary) (parasternal) nodes—located along the internal thoracic vessels—drain the medial part of the breast, anterior parietal pleura, and thoracic wall*

i. **retrosternal**—the lower portion of the internal thoracic (parasternal) chain—drain the anterior diaphragm and the anterior abdominal wall above the umbilicus

2. *Right tracheobronchial nodes—located along the trachea and bronchi—drain the trachea, bronchi, heart, and lungs*

i. bronchopulmonary—at the hilum of the lung

a. pulmonary—within the lungs

ii. tracheal—located along the trachea and upper esophagus

3. *Right anterior mediastinal nodes—located anterior to the aortic arch—drain the right anterior abdominal wall above the umbilicus*

C. Right Subclavian Trunk (169)

1. *Axillary nodes*

i. **apical group**—receives other axillary lymphatics. The apical group also receives efferents from the infraclavicular (subclavicular) nodes which drain the upper portion of the breast. The apical group is located along the medial aspect of the axillary vein and behind the pectoralis minor muscle.

ii. lower group—drains into the apical group

a. **central group**—receives some lymph from other axillary lymphatics—empties into the apical lymphatics—located along the axillary artery in the axillary apex

b. **lateral group**—drains portions of the upper limb

1) cubital (supratrochlear) (epitrochlear)—located above the medial epicondyle of the humerus

c. **anterior (pectoral) group**—drains the anterior chest, abdominal wall above the umbilicus, and lateral parts of the breast—located in the anteromedial axillary wall along the inferior border of the pectoralis minor muscle

d. **posterior (subscapular) group**—located under the posterior axillary fold—drains the posterior thoracic wall above the iliac crests

III. THORACIC DUCT (249)—receives all lymphatics below the diaphragm and the left half of the thorax. It begins at the cisterna chyli at the level of L1–L2 and usually ends in the left brachiocephalic vein.

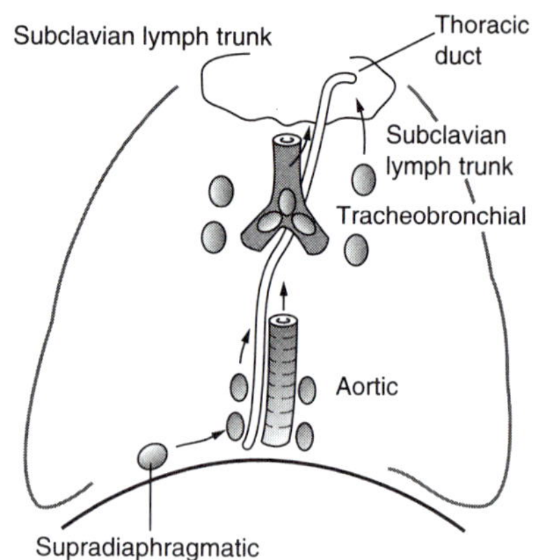

A. Left Jugular Trunk

B. Left Bronchomediastinal Nodes—main tributaries as on the right

C. Left Subclavian Nodes

D. Left Anterior Mediastinals

E. Posterior Mediastinals—drain the esophagus and posterior diaphragm—located near the lower esophagus

F. Upper Intercostals—located between the ribs and near the vertebrae—drain portions of the posterior abdominal wall

IV. CISTERNA CHYLI (249)—located at the level of L1–L2 and to the right of the aorta

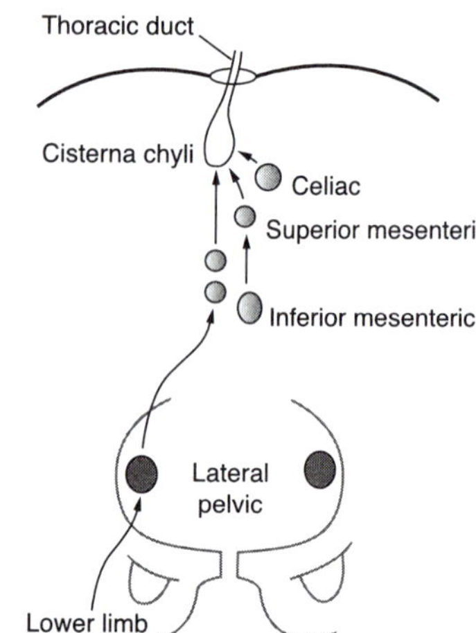

A. Lower Posterior Intercostals

B. Lumbar Nodes—located on the posterior abdominal wall—drain portions of the abdominal wall below the umbilicus.
(The lumbar nodes are sometimes described as those above the aortics. The terms "lumbar nodes" and "aortic nodes" are sometimes used interchangeably.)

C. Aortic Nodes—drain suprarenals, upper ureter, testes, ovaries, tubes, fundus of the uterus, and the posterior abdominal wall
NOTE: Lymph nodes for the abdominal portion of the intestinal tract are sometimes classified as:
Central Nodes—nodes located near the origins of the celiac, superior mesenteric, or inferior mesenteric arteries
Regional Nodes—nodes located along the major branches of the arteries, usually within the mesentery
Local Nodes
In addition, lymphatic follicles may be located within the walls of the intestine (e.g., Peyer's patches).

1. *Renal nodes—drain the kidneys and upper ureter*

2. *Celiac nodes*

   i. left gastric—lesser curvature of the stomach and abdominal part of the esophagus

   ii. right gastric—drain the lesser curvature

   iii. right gastro-omental—drain the pylorus and greater curvature

iv. pancreaticosplenic—drain the greater curvature of the stomach plus the spleen, and the pancreas

v. pancreaticoduodenal

vi. hepatic—liver and gallbladder

vii. gastroduodenal—1st and 2nd parts of the duodenum. (Lymph nodes along the gastroduodenal artery are sometimes referred to as the pyloric nodes.)

viii. pyloric—drain the pylorus and 1st part of the duodenum

3. *Superior mesenteric nodes—3rd and 4th parts of the duodenum, pancreas, jejunum, ileum, ascending colon, and proximal transverse colon*

4. *Inferior mesenteric nodes—distal transverse colon, descending colon, and upper ½ of the rectum. They also receive some efferents from the sacral nodes.*

5. *Common iliacs—drain the middle of the ureter*

   i. **external iliacs**—bladder, vas deferens, lower ureter, cervix, prostate, and middle ⅓ of the vagina. They also have efferents from the inguinal nodes.

      a. superficial inguinal—located near the termination of the great saphenous vein—drain the abdominal wall below the umbilicus, lower ⅓ of the vagina, vulva, anal canal below the pectinate line, superficial tissues of the perineum, gluteal region, the body of the penis, and the lower limb

      b. deep inguinal—located in the femoral triangle—drain the clitoris, glans penis, and lower limb. They are associated with blood vessels (particularly the femoral artery). The highest deep inguinal node located in the femoral triangle is called "**Cloquet's node.**"

         1) popliteal nodes—drain the leg and foot—located in the popliteal fossa

         2) anterior tibial nodes—located in the anterior compartment

   ii. **internal iliacs**—lower ureter, middle and upper ⅓ of the vagina, body of the uterus, prostate, rectum, anus (above the pectinate line), and deep portions of the perineum and gluteal region

      a. sacral—cervix, lower ½ of the rectum, anus above the pectinate line, prostate, and upper ⅓ of the vagina

Nervous Systems

I. CENTRAL NERVOUS SYSTEM (CNS)

A. Composition—brain and spinal cord

B. Tract—a collection of axons in the CNS

C. Nucleus—a collection of nerve cell bodies inside the brain or spinal cord

D. Synapse—attachment between neurons

II. PERIPHERAL NERVOUS SYSTEM

A. Composition

1. *Cranial nerves*
2. *Spinal nerves*

B. Nerves—a group of axons

C. Neurons—nerve cells

1. *Axons*
   - i. axon hillock—origin of the axon
   - ii. axon terminal—the end of the axon—for synapses with other nerve cells
2. *Dendrites—extensions of the cell which increase its synaptic surface*

D. Ganglion—a collection of nerve cell bodies

E. Plexus—a collection of nerves

F. Synapse—attachment between neuron and an organ

G. Roots

1. *Dorsal—afferent (sensory) fibers from spinal ganglia*
2. *Ventral—efferent (motor) fibers from spinal cord to spinal nerves*

H. Spinal Nerves—the union of dorsal and ventral roots which innervate a body segment

1. *Dorsal ramus—nerve fibers to the back*
2. *Ventral ramus—nerve fibers to the limbs and anterior body wall*
3. *Pairs—Cervical 8—Thoracic 12—Lumbar 5—Sacral 5—Coccygeal 1*

III. AUTONOMIC NERVOUS SYSTEM (ANS) (152–154, 306)

A. Sympathetic Nervous System (SNS)—has vasoconstrictor, pilomotor, and sudomotor fibers

1. *Preganglionic—from brain or spinal cord to a ganglion. After entering the thoracic and lumbar (L1 or L2) ganglia, they end (synapse) in paravertebral ganglia or exit the paravertebral ganglia to end in prevertebral ganglia or exit the prevertebral ganglia to end in a plexus.*
   NOTE: *Preganglionic fibers are white (myelinated). The T1-L2 cord segments are attached to a paravertebral ganglion via a white ramus. These are the* ***only*** *spinal cord segments that contain preganglionic sympathetic cell bodies. The preganglionic neurotransmitter is acetylcholine.*

2. *Postganglionic—from ganglion to target tissue. The neurotransmitter is usually norepinephrine.*
   - i. paravertebral ganglia (sympathetic trunk)—located alongside the vertebrae **or**
   - ii. prevertebral ganglia—located in front of the vertebrae

3. *Examples of some sympathetic neuronal linkages*
   - i. supplying innervation for the body wall, limbs, and cervical and thoracic areas
     - a. The preganglionic neurons originate in the spinal cord.
     - b. The postganglionic neurons originate in the paravertebral ganglion and thence travel to target organs.
   - ii. for the abdomen
     - a. The preganglionic neurons originate in the spinal cord and pass thru the paravertebral ganglia, where they form splanchnic nerves (greater, lesser, and least) and descend into the abdomen. These neurons have passed thru paravertebral ganglia without synapsing.
     - b. The postganglionic neurons originate in prevertebral ganglia (e.g., celiac, preaortic, or mesenteric) and are distributed to the viscera along blood vessels.
   - iii. in the pelvis
     - a. The preganglionic neurons originate in the spinal cord.
     - b. The postganglionic neurons originate in a hypogastric plexus or lumbar or sacral sympathetic trunk and thence to pelvic viscera.

B. Parasympathetic Nervous System (PNS)

1. *Cranial portion—via cranial nerves III, VII, IX, and X*

2. *Sacral portion—connects to the spinal cord via spinal nerves S2, S3, and S4*

3. *Examples of parasympathetic neuronal linkages*

   i. in the head

      a. Preganglionic parasympathetic cranial neurons originate in the brain.

      b. Postganglionic neurons originate in ganglia.

         1) CN III—ciliary ganglion to sphincter pupillae and ciliary muscles

         2) CN VII—submandibular ganglion to submandibular and sublingual salivary glands

         3) CN IX—otic ganglion to the parotid gland

   ii. in the thorax

      a. Preganglionic neurons originate in the brain.

      b. Postganglionic neurons originate in a plexus associated with the great vessels and thence travel to the nearby heart and lungs.

   iii. in the abdomen—as far as the splenic flexure

      a. Preganglionic neurons originate in the vagal nucleus.

      b. Postganglionic neurons originate in ganglia in the walls of the viscera.

   iv. in the abdomen—from the splenic flexure onward

      a. Preganglionic neurons originate in the sacral spinal cord (S2–S4).

      b. Postganglionic neurons originate in the walls of the viscera (having passed thru the hypogastric plexus).

   v. in the pelvis

      a. Preganglionic neurons originate in the sacral spinal cord (S2–S4).

      b. Postganglionic neurons originate in the inferior hypogastric plexus or in the walls of the viscera.

      c. The pre- and postganglionic neurotransmitter is acetylcholine.

IV. VISCERAL AFFERENT NERVES—usually considered part of the ANS (306, 307)

A. Share Pathways with the ANS

1. *In the cervical region—travel along with cervical splanchnic nerves to the sympathetic chain and thence down to white rami in the thorax to enter the thoracic spinal cord*

2. *In the thorax—travel along with splanchnic nerves to the sympathetic chain and thence to the thoracic spinal cord via white rami*

3. *In the abdomen*

   i. foregut derivatives via celiac plexus and greater splanchnic nerves to T5–T9

   ii. midgut derivatives via superior mesenteric plexus and lesser splanchnic nerves to T10 and T11

   iii. hindgut derivatives via aortic plexus and lumbar splanchnics to L1 and L2

4. *In the pelvis*

   i. along sympathetic pathways to lumbar splanchnics to lumbar spinal nerves

   ii. along parasympathetic pathways to S2–S4

B. Characteristics of Visceral Afferents

1. *Those accompanying sympathetics carry pain fibers.*

2. *Those accompanying parasympathetics carry pressure sensations which may ultimately result in changes in the smooth muscle (such as peristalsis).*

3. *Noxious stimuli and distension travel via visceral afferent fibers to the nodose ganglion and thence to the CNS.*

4. *Are the basis of referred pain*

V. ENTERIC NERVOUS SYSTEM (ENS)—composed of neurons derived from the neural crest which have migrated along the Vagus (CN X) nerve to the intestinal tract. Intestinal motility and the peristaltic reflex are ENS functions. Motility consists of segmental, nonpropulsive mixing and peristaltic, propulsive movements. The ENS organizes the mixing movements to bring about the peristalsis. The ENS is frequently considered part of the ANS.

TABLE 1-1. SYMPATHETIC FIBERS

PREGANGLIONIC FIBERS (Cell Bodies in Spinal Cord from T1 to L2 or L3)	**POSTGANGLIONIC FIBERS** (Cell Bodies Mostly in Ganglia Close to Vertebral Column)	**TARGET ORGANS**
	In the Limbs, Body Wall, and Skin of the Neck	
Preganglionic fibers via ventral root and white ramus to a paravertebral ganglion	Paravertebral ganglia send postganglionic fibers. They pass thru a gray ramus and are carried by spinal nerves to the target organs listed on the right.	Smooth muscle of blood vessels (constriction) Muscles of hair follicles (contraction) Sweat glands (secretion)
	In the Head	
Preganglionics from T1 or below	Sympathetic cervical ganglia. Postganglionics are distributed along the carotid arteries	Carotid vessels
	In the Neck	
	Cervical and upper five thoracic ganglia	Heart—via the cardiac plexus (increases beat and contractions and dilates coronary arteries) Lungs—via the pulmonary plexus (dilates bronchi)
	In the Abdomen	
Greater splanchnic nerves (from paravertebral ganglia T5–T9)	Celiac ganglion **NOTE:** Splanchnic nerves transmit preganglionic fibers from paravertebral ganglia (the sympathetic chain) to prevertebral ganglia and plexuses	Stomach (inhibits gastric juices, decreases motility, and contracts pyloric sphincter) Pancreas (decreases enzyme secretion) Liver (increases glycogenolysis and decreases bile) Spleen (contracts, causing increase of blood into the circulation) 1st and 2nd parts of duodenum
Greater, lesser, and least (lowest) splanchnic nerves	Superior mesenteric plexus	Midgut derivatives 2nd and 3rd parts of the duodenum Jejunum and ileum Proximal ⅔ of colon (decreases motility)
Lesser and least (lowest) splanchnic nerves (T10–T12)	Renal plexus	Kidney (constricts vessels and decreases urine production) Upper ureters
Splanchnic nerves	Celiac ganglia Suprarenal medulla	Blood vessels in the suprarenal Suprarenal chromaffin cells (increases catecholamines)
First and second lumbar splanchnic nerves	Lumbar sympathetic trunk Superior hypogastric plexus	Descending colon Sigmoid colon Lower ureters
	In the Pelvis	
Sympathetic chain	Inferior hypogastric plexus	Bladder (relaxes detrusor and contracts sphincter, causing retention of urine) Uterus (inhibits contraction) Ejaculatory ducts, seminal vesicles, and prostate (constriction resulting in ejaculation)

TABLE 1-2. CRANIAL AND CAUDAL PARASYMPATHETIC FIBERS

PREGANGLIONIC FIBERS (Cell Bodies in Brain or Sacral Cord)	**POSTGANGLIONIC FIBERS** (Cell Bodies in Ganglia or Viscera)	**TARGET ORGANS**
Oculomotor (CN III)	Ciliary ganglia—located deep in the orbit	Sphincter pupillae (constricts pupil) Sphincter ciliaris (accommodates lens for near vision)
Facial (CN VII)	Pterygopalatine ganglia—located in its fossa	Lacrimal gland (promotes tearing) Mucous glands of nose, palate, and pharynx
	Submandibular ganglia—located in its fossa	Salivary glands (stimulates saliva)
Glossopharyngeal (CN IX)	Otic ganglia—located in the infratemporal fossa	Parotid gland (secretes parotid saliva)
Vagus (CN X)	**In the neck**	Larynx
	In the thorax	Lungs (constricts bronchi) Heart (decreases contractions, slows heart rate, constricts coronaries)
	In the abdomen (includes ENS components)	Stomach (stimulates gastric juices) Liver (stimulates bile and increases gluconeogenesis) Pancreas (increases endocrine secretions) Intestines (increases peristalsis as far as the proximal ⅔ of the colon) Upper ureters
Pelvic splanchnics (nervi erigentes) (S2, S3, S4)	**In the abdomen**	Colon (increases peristalsis in distal ⅓) Sigmoid colon and rectum Lower ureters
	In the pelvis	
	Bladder wall	Contracts bladder (detrusor) muscles
	Bladder sphincter	Relaxes sphinctor
	In the perineum	
	Penis and clitoris	Dilates blood vessels
	Bulbourethral glands—male	Increases secretion
	Vestibular glands—female	Increases secretion

Vagus

Vagus fibers (70, 307) supply some of the structures in the neck, thorax (including the heart and lungs), and abdomen (the major part of the digestive system).

I. IN THE HEAD (119)—The Vagus originates in the medulla and exits the cranium via the jugular foramen along with CN IX and CN XI.

A. Forms the Superior Jugular (Vagal) Ganglion (120)

1. *Meningeal branch*
2. *Auricular branch—sensory branches to tympanic membrane and auricle*

II. IN THE NECK (129)—The Vagus forms the inferior (nodose) (vagal) ganglion. It then descends between the internal carotid artery and the internal jugular vein and enters the carotid sheath between the common carotid artery and the internal jugular vein.

A. Pharyngeal Branch (119)—also contains motor fibers from the Accessory nerve (CN XI)—has motor fibers to the soft palate and contributes to the pharyngeal plexus

B. Carotid Sinus Nerve (119)—along with CN IX, supplies the carotid body

C. Superior Laryngeal Nerve (70)—located behind the internal carotid artery. **The superior laryngeal nerve inhibits the "immersion reflex" which brings about diversion of blood to the brain, heart, and adrenals.**

1. *Internal laryngeal nerve—sensory*
   i. supplies mucosa of the larynx above the vocal cords
2. *External laryngeal nerve—lies close to the superior thyroid artery* ***(a surgical danger area).***

   The nerve can be injured during thyroidectomy and produce permanent voice changes, such as hoarseness or a change in pitch.

   i. supplies the cricothyroid muscles (which increase tension on the vocal cords and therefore increase pitch)

D. Right Recurrent Laryngeal Nerve (70)—arises from the Vagus where the Vagus crosses the subclavian artery. It hooks around (backward and upward) the subclavian artery to ascend between the trachea and esophagus. It may lie behind, in front of, or around the inferior thyroid artery **(surgical danger areas).**

1. *Supplies all the muscles of the larynx except the cricothyroid muscle*
2. *Supplies the mucous membranes of the larynx (below the vocal cords) and the upper trachea*

E. Cardiac Branches—to the cardiac plexus, where it joins branches from the sympathetic trunk

III. IN THE THORAX (228)

A. The **Vagus nerves**—Pulmonary branches go to the pulmonary plexus, where they synapse and supply bronchial constrictor muscles. After leaving the pulmonary plexus, the Vagus nerves form the esophageal plexus.

B. The **right Vagus** enters the thorax on the right side of the trachea. It reaches the thorax behind the right brachiocephalic vein. It then proceeds to the right pulmonary plexus and continues on to the esophageal plexus.

C. The **left Vagus** descends to the aortic arch, hooks around it, and gives off the left recurrent laryngeal nerve. The left Vagus then continues on to the pulmonary and esophageal plexuses.

➢ 1. *Left recurrent laryngeal nerve (70)—arises from the Vagus as it crosses the aorta. It hooks behind ligamentum arteriosum (former ductus arteriosus). In the neck, it ascends upward between the trachea and the esophagus. The left recurrent laryngeal nerve lies behind, in front of, or around the inferior thyroid artery **(surgical danger areas)**.*

i. supplies the muscles of the larynx except the cricothyroid

ii. supplies the mucous membranes of the larynx (below the vocal cords) and the upper trachea

D. After leaving the esophageal plexus, the Vagus nerves form the anterior and posterior vagal trunks.

➢ IV. IN THE ABDOMEN (306)—The Vagus nerves enter the abdomen on the esophagus to supply derivatives of the foregut and midgut.

A. Posterior Vagal Trunk—extends to the greater omentum (as the posterior gastric)

1. *Sends branches to posterior surface of stomach, pancreas, spleen, celiac plexus, and superior mesenteric plexus*

B. Anterior Vagal Trunk—extends to the lesser curvature of the stomach (as the anterior gastric)

1. *Supplies anterior surface of stomach, pylorus, and liver*

Back

➢ **I. VERTEBRAE (142)—composed of a body and a neural arch**

A. Cervical (7)—flexion, extension, and rotation

B. Thoracic (12)—rotation, some flexion, and extension

C. Lumbar (5)—flexion and extension

D. Sacral—composed of 5 fused vertebrae

E. Coccygeal

II. VERTEBRAL BODY—progressively larger inferiorly.

The body has a nutrient foramen on the anterior surface and a foramen for the basivertebral vein on the posterior surface.

➢ **III. VERTEBRAL (NEURAL) ARCH (143, 144)—protects the cord.**

A diving accident may result in a compression fracture of C1, breaking the arch into pieces and pushing them outward.

A. Pedicles—extend posteriorly from the body to form laminae

1. *Superior and inferior vertebral notches*

i. formed by indentations of the pedicles

ii. The inferior intervertebral notches of a superior vertebra "articulate" with the superior vertebral notches of an adjacent inferior vertebra to form right and left **intervertebral foramina** thru which spinal nerves and blood vessels to the cord pass.

B. Laminae of the Vertebral Arch—located between the pedicles and the spinous process. They meet in the midline to form the spinous process.

➢ **IV. VERTEBRAL PROCESSES (143, 144)**

A. Transverse Processes—located at the junction of pedicle and lamina. They project laterally between the superior and inferior facets and are the attachment for deep muscles.

B. Superior and Inferior Articular Processes—located at the junction of the pedicle and lamina. They have facets for adjacent vertebrae.

V. FORAMEN (143, 144)

A. Vertebral (Spinal) Foramen—foramina form the vertebral canal

1. *Formed by the vertebral body, pedicle, and lamina*
2. *Contains meninges*
   - i. dura—from the foramen magnum to S2
   - ii. arachnoid (the subdural space is between the arachnoid and the dura while the subarachnoid space is between the pia and arachnoid)
   - iii. pia—a transparent vascular covering of the cord

B. Nutrient (Arterial) Foramen—located anteriorly on the body

C. Basivertebral—located posteriorly on the body and transmits the basivertebral vein

D. Intervertebral Foramen—for spinal nerves and blood vessels

E. Sacral Foramina—for the sacral nerves

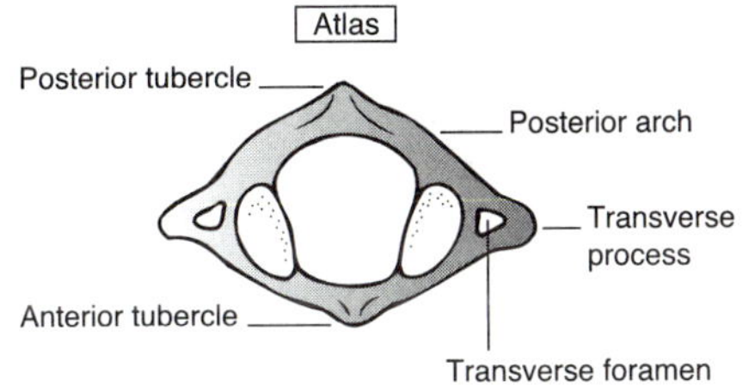

VI. SPECIAL FEATURES

A. Cervical (12, 13)—contain transverse foramina for the passage of the vertebral artery

1. *Atypical cervical vertebrae*
   - i. C1 (atlas)—**no spinous process or body** (replaced with a posterior and an anterior tubercle, respectively)—has an articular facet on the central posterior border of its anterior arch for articulation with the dens of C2. The superior articular facets are concave and articulate with the occipital condyles.
   - ii. C2 (axis)—has a dens (odontoid process) which is a protuberance that projects superiorly thru C1's vertebral foramen and articulates with its facet. The dens takes the place of the C1 vertebral body. It extends into the atlas and permits rotation. The dens and the C1 vertebra act as a single functional unit.
2. *Typical cervical vertebrae—contain transverse foramina*
   - i. C3–C6—the spinous processes are bifid
   - ii. C7—not bifid—has a prominent spinous process

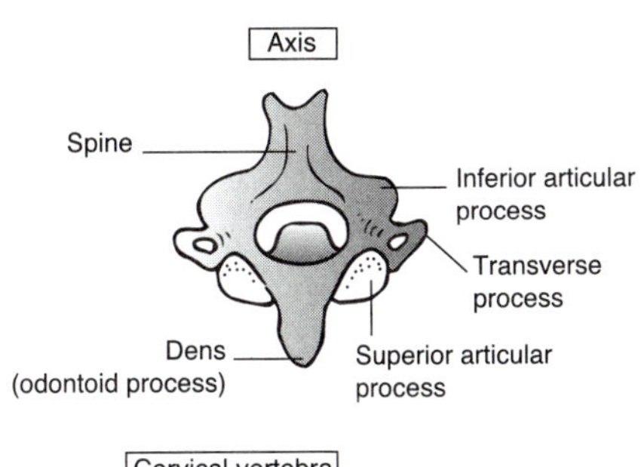

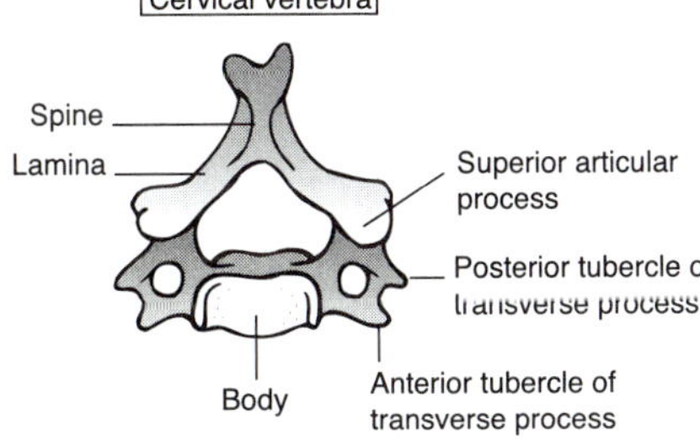

B. Thoracic (143)—increase in size from above downward

1. *Typical—T5–T8—have a long spinous process*
2. *Atypical—T9–T12—have tubercles*

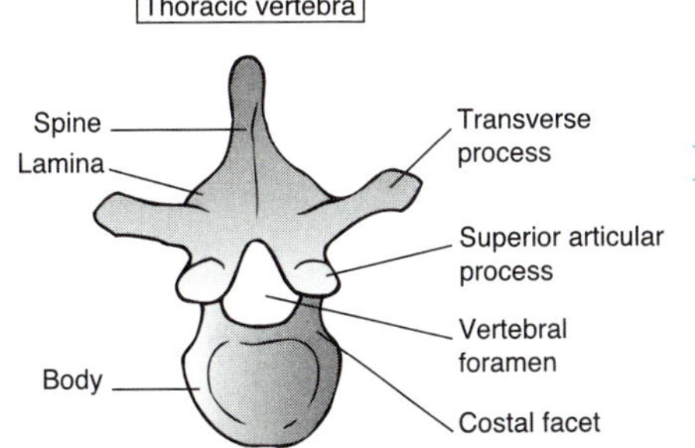

3. *Costal facets—The upper and lower costal facets are located between the pedicle and body. The transverse costal facet is on the transverse process for T1–T12.*

C. Lumbar (144, 146)—large bodies and short horizontal spinous processes. The pedicles have deep inferior intervertebral notches.

Spina bifida involves the incomplete formation of the lamina (posterior vertebral arches) at L4(5) to S1.

Sciatica (pain along the sciatic nerve distribution) may result from a disk herniation.

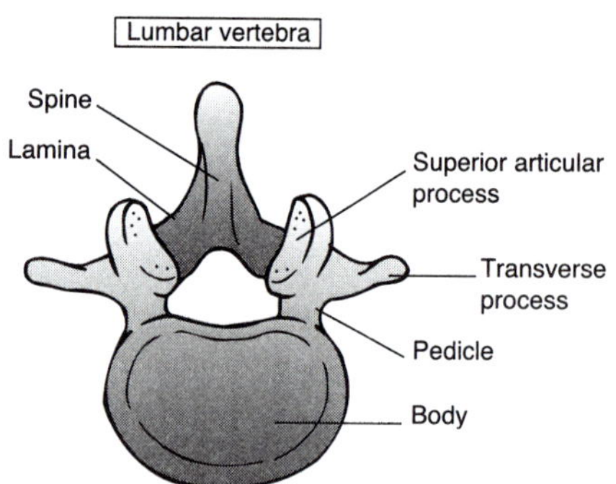

D. Sacral—have only 4 pairs of foramina

1. *Ala—one on either side of the body*

2. *Sacral hiatus—between the sacral cornua—located at S2–S3*

E. Coccyx—connected to the sacrum by a ligament

VII. ARTICULATIONS OF TYPICAL VERTEBRAE

A. Intervertebral Disks (144)—from C2 to L5

1. *Composed of nucleus pulposus (gelatinous) surrounded by the annulus fibrosis (fibrous component) surrounded by a hyaline component. The disks act as a cushion between the vertebral bodies and are thickest in the lumbar region. The longitudinal ligaments help support the annulus.*

A herniated disk compresses the adjacent inferior spinal nerve roots. This may affect cutaneous sensation, visceral sensation, and muscle strength.

B. Synovial Joints (172)

1. *Costovertebral joints—between the rib and vertebral body facet*

2. *Costotransverse joints—between the rib and transverse process*

VIII. LIGAMENTS (14–16, 146, 147)

A. Atlanto-occipital—from the foramen magnum to the atlas

B. Cruciate (Cruciform) Ligament of the Atlas—stabilizes the dens (holds it to the atlas) and the joint between atlas and axis

C. Supraspinous—connect tips of spinous processes (C7 to sacrum). It prevents severe flexion.

D. Ligamentum Nuchae—from the occipital protuberance to the cervical vertebrae and their spinous processes (C2–C7). It is a continuation of the supraspinous ligament in the neck.

E. Interspinous—between spinous processes

F. Longitudinal Ligaments—connect bodies and disks

1. *Anterior—supports the intervertebral disk laterally and resists vertebral extension*

   It is strained or torn in a "whiplash" injury.

2. *Posterior—reinforces the anterior aspect of the vertebral canal and prevents severe flexion*

   Herniated disks go posterolaterally.

G. Flaval (Ligamentum Flavum)—elastic tissue between laminae

IX. POSTERIOR CERVICAL MUSCLES (160, 164, 165)

A. Rectus Capitis (Short) Muscles—**from** the occipital bone

1. *Anterior—to the atlas—flexes and rotates the head*
2. *Lateral—to the atlas—flexes the head*
3. *Posterior—to the axis—extends and rotates the head*

B. Long Muscles—**from** one or more cervical vertebrae

1. *Longus muscles*
   i. longus colli—to thoracic vertebrae—flexes the cervical vertebrae
   ii. longus capitis—to the occipital bone—flexes the head and cervical vertebrae
2. *Scalene muscles*
   i. scalenus anterior and medius—elevate the 1st rib
   ii. scalenus posterior—elevates the 2nd rib
3. *Splenius muscles—extensors of the spine*
   i. splenius capitis—to the mastoid—located between the spinous process and mastoid
   ii. splenius cervicis—to upper thoracic vertebrae—located between a spinous process and transverse process above

C. Sternocleidomastoid—moves the head but is not attached to the vertebral column

X. VERTEBRAL EXTENSOR GROUP—ERECTOR SPINAE—(Transversocostal) (161)—superficial muscle group (but deep to the back muscles which extend from the back to the upper limb). (Its longitudinal divisions are called cervicis, thoracis, and lumborum.)

Contraction of the erector spinae muscles on both sides extends the back (vertebral column), while contraction of one side produces lateral flexion.

A. Iliocostalis—lateral fibers—ribs to transverse processes

B. Longissimus—middle fibers—spines to ribs

C. Spinalis—most medial fibers—spines to spines

➢ **XI. EXTENSOR AND ROTATOR GROUP (Transversospinal) (162)—the deep muscle group—located in vertebral column grooves. Contraction of both sides extends the back (vertebral column), while contraction of one side rotates the back.**

A. Long rotators (semispinalis cervicis and thoracis)—from transverse process to transverse process above

B. Short rotators (rotatores and multifidis)—from vertebral process to spinous process above

XII. ABDOMINAL MUSCLES WHICH FLEX THE SPINE

A. Posterior Abdominal Wall Muscles—psoas major and minor, quadratus lumborum

B. Anterior Abdominal Wall Muscle—rectus abdominis

➢ **XIII. VEINS (159)—consist of vertebral plexuses**

A. Spinal—drain into the intervertebral veins

B. Vertebral—external vertebral plexus drains into azygos veins. The basivertebral veins drain into anterior internal vertebral (or epidural) venous plexus and thence into the intervertebral vein.

C. The Vertebral Venous Plexuses—alternate routes for blood to flow from the pelvis to the superior vena cava.

➢ **XIV. SPINAL CORD (148, 149, 151, 155, 156)**

A. The Spinal Cord begins at the foramen magnum and ends at the conus medullaris. However, the spinal cord proper ends at the level of L2. The dural sac ends at approximately the S2 level.

B. End of Conus Medullaris—at birth L2 or L3—adult L1 or L2

C. Filum Terminale—a continuation of the conus medullaris to the coccyx. It descends thru the center of the cauda equina.

D. Cauda Equina—consists of lumbar and sacral spinal nerve rootlets inferior to the termination of the spinal cord

E. Fissures—anterior and posterior median fissure

F. Spinal Meninges and Spaces (155)

1. *Epidural space—between dura and bone—contains connective tissue, fat, and veins (vertebral venous plexus)*
2. *Dura—from the foramen magnum to the level of S2*
3. *Subdural space—between the dura and arachnoid*
4. *Arachnoid—covers the cauda equina*
5. *Subarachnoid space—between pia and arachnoid—surrounds the cord and contains the cerebrospinal fluid (CSF)—**location for spinal anesthesia***
6. *Pia—applied to the cord. The pia covers the filum terminale.*

A lumbar puncture (spinal tap) is usually done at L3–L4 or L4–L5 because the spinal cord ends at L2. The crest of the ilium is at vertebral level L4. The needle extends thru the skin, fascia, supraspinous ligament, intraspinous ligament, epidural space (the needle stops here for epidural anesthesia), dura, and arachnoid and then stops in the subarachnoid space.

G. Arteries to the Spinal Cord (157, 158)

1. *Spinal Arteries (from the vertebrals)—supply the cord—have one anterior and two posterior branches—go thru their respective rami*
2. *Radicular (Segmental Medullary) Arteries—come from the intercostal, lumbar, and other arteries and are the major blood supply to thoracic and lumbar cord*
   i. The great anterior radicular branch supplies the inferior spinal cord.

H. Nerves

1. *Cell bodies and roots*
   i. ventral horn and ventral root—motor fibers
   ii. dorsal horn and dorsal root sensory fibers
2. *Spinal nerves (31 pairs)—sensory and motor fibers—formed outside of the dura. Dorsal primary rami innervate the back muscles.*

Head

Chapter 2

Page numbers in black. Netter plate numbers in green.

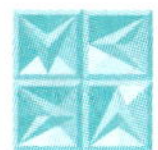

External Carotid Artery

The external carotid artery supplies muscular, oral, and visceral portions of the head and neck (except the brain and orbit).

I. COURSE (29, 63)—The external carotid artery arises from the common carotid near the upper border of the thyroid cartilage and courses upward toward the angle of the jaw at the back of the mandible. It then courses under the stylohyoid muscle to pierce the parotid gland. It ends behind the neck of the mandible. The superficial temporal and maxillary are its terminal branches.

II. BRANCHES (35, 63, 130, 131)

A. Superior Thyroid (130)

1. *Thyroid—to the superior portion of the thyroid*
2. *Muscular—to the anterior neck muscles*
3. *Superior laryngeal—to the larynx*

B. Lingual (53)—to the tongue and muscles attached to the hyoid bone

C. Facial (External Maxillary) (35)—The chief artery of the face arises in the carotid triangle and extends upward to the angle of the mandible, continues on to the submandibular gland and then downward and forward to the lower border of the mandible, which it crosses, and then turns upward toward the face. The facial artery supplies the submental area, palate, tonsils, submandibular gland, lips, lip muscles, and nasal skin.

1. *Branches in the neck*
   - i. ascending palatine—to the upper pharynx, soft palate, and tonsil
   - ii. tonsillar—to the upper pharyngeal constrictor muscles
   - iii. submandibular—to the submandibular gland
   - iv. submental—to the submandibular and sublingual glands
2. *Branches in the face*
   - i. superior labial—to the upper lip and nasal septum
   - ii. inferior labial—to the lower lip
   - iii. nasal
   - iv. muscular—to the pterygoid, masseter, and buccinator muscles

D. Occipital (17, 95, 164)

1. *Muscular—to the posterior neck muscles*

2. *Auricular—enters the mastoid foramen to supply the dura*

3. *Meningeal—thru the jugular foramen to supply the posterior fossa dura*

4. *Descending branch—to the posterior neck muscles*

E. Posterior Auricular (35)—thru the stylomastoid foramen to the tympanic area, semicircular canals, and posterior scalp

1. *Stylomastoid branch—supplies the labyrinth of the inner ear*

F. Ascending Pharyngeal (35, 130)—to the pharyngeal muscles, soft palate, tonsils, and meninges (via the jugular foramen)

G. Superficial Temporal (35)—to the parotid gland (which it passes thru), auricle, forehead, and temporal and parietal skin

1. *Transverse facial—to the parotid and overlying skin*

2. *Anterior terminal branches to upper superficial muscles and skin*

H. Maxillary (35, 63)—multiple branches including those to the mandible, teeth, muscles of mastication, deep portions of the face, auricle, palate, orbit, and calvaria. The maxillary artery begins within the parotid gland posterior to the neck of the mandible.

1. *First part of the maxillary (mandibular) artery (35, 131)—supplies areas such as the skull, dura, tympanic membrane, lower teeth, and temporomandibular (TM) joint*

   i. deep auricular—to the TM joint, external acoustic (auditory) meatus, and outer tympanic membrane

   ii. anterior tympanic—to the tympanic membrane

   iii. middle meningeal (95)—to the dura and interior of calvaria—enters foramen spinosum to the middle cranial fossa to course along the greater wing of the sphenoid

      a. superior tympanic—to the tensor tympani

      b. anterior branch—along the greater wing of the sphenoid and thence along the parietal bone between dura and cranium

      c. posterior branch—along the greater wing of the sphenoid and thence along the temporal bone between dura and cranium

iv. accessory meningeal—enters the foramen ovale to supply the dura mater

v. inferior alveolar (dental) (35)—supplies buccal mucous membrane and teeth, then passes thru the mandibular canal and foramen to give off the:

a. mental artery to the mandible

2. *Second part of maxillary (pterygoid) artery (63)—supplies the mandible and the buccinator, masseter, and temporal muscles*

i. branches to the masseter and temporal muscles

ii. buccal artery (35, 63)—to the buccinator muscle and mucous membrane of the cheek

3. *Third part of the maxillary (pterygopalatine) artery (generally follows the distribution of the Maxillary nerve)—supplies areas such as the palate, nasal cavity, maxilla, and the lower portion of the orbit and eyelid*

i. alveolar (dental) branches (63)—supply molars, premolars, and the lining of the maxillary sinus

ii. infraorbital (35)—to canine and incisor teeth, inferior oblique and inferior rectus muscles, lacrimal gland and sac, lower eyelid, part of the maxilla, and upper lip

iii. descending (greater) palatine (63)—to the palate, palatine, and maxillary bones

iv. artery of the pterygoid canal—to the upper pharynx

v. pharyngeal—supplies upper mucous lining of nose, pharynx, and sphenoidal sinus

vi. sphenopalatine (63)—thru the sphenopalatine foramen to the posterolateral wall of the nasal cavity and the sinuses which open into it (frontal, maxillary, ethmoid, and sphenoid)

Internal Carotid Artery

The internal carotids are the arterial supply to the eye, forehead, nose, and anterior part of the brain.

I. COURSE (63, 131)

A. Cervical Part

1. *The artery begins at the bifurcation of the common carotid, opposite the superior aspect of the thyroid gland, and runs in front of the transverse processes of the cervical vertebrae.*
2. *The carotid sinus is a dilation located at the origin of the internal carotid.*
3. *The internal carotid artery lies generally behind the external carotid and under the sternocleidomastoid muscle. It passes beneath the parotid gland.*
4. *As the internal carotid artery ascends, it goes behind the styloid process, the Glossopharyngeal nerve, and the pharyngeal branch of the Vagus. Behind it is the sympathetic trunk, while lateral to it are the jugular vein and Vagus nerve.*

B. Petrous Part

1. *The internal carotid artery passes thru the foramen lacerum and thence thru the carotid canal in the temporal bone at the base of the skull.*
2. *The internal carotid artery courses forward, then leaves the canal to enter the skull to become the:*

C. Cavernous Part—passes forward and upward along the clinoid process and perforates the dura

D. Cerebral Part—passes lateral to the optic chiasma. The terminal branches of the internal carotid are the anterior and middle cerebral arteries. The anterior cerebral passes over the optic nerve. The middle cerebral passes upward to the lateral cerebral (Sylvian) fissure to divide into cortical branches.

II. BRANCHES (130–135, 141)

A. Cervical Part—no branches

B. Petrous Part—branches to tympanic cavity and pterygoid canal

C. Cavernous Branches—to the hypophysis and cavernous sinuses

1. *Hypophyseal*
2. *Anterior meningeal branch—to the anterior dura*

3. *Ophthalmic artery—to the orbit, eye (via the optic canal), forehead (via the supraorbital fossa), and ethmoid*
   i. retinal (central artery of the retina)
   ii. ciliary—to eyeball
   iii. muscular—to orbital muscles and iris
   iv. lacrimal—to lacrimal gland
   v. supraorbital—to forehead via supraorbital notch
   vi. palpebral—to eyelids
   vii. meningeal—to meninges via the supraorbital foramen
   viii. ethmoidal—to sinuses

D. Cerebral Part

1. *Anterior cerebral—to the anterior portion of the cerebral hemisphere*
   i. supplies part of the frontal and parietal lobes (the medial surfaces of the cerebral hemispheres)
   ii. supplies the corpus callosum and septum pellucidum
   iii. connects with the opposite anterior cerebral (via the anterior communicating artery)
2. *Middle cerebral—to the lateral surfaces of the cerebral hemispheres—supplies part of the frontal and parietal lobes*
3. *Posterior communicating—connects to the posterior cerebral (a branch of the basilar). The basilar is a "joining" of the vertebral arteries.*

III. CAROTID INJURIES—The carotids are subject to aneurysm, blunt trauma, and complications of endarterectomy.

A. **Leakage, hematoma, or AV fistula formation**

B. **Damage to the intima—dissection, pseudoaneurysm formation, thrombosis, and stroke**

C. **Neurologic and muscle deficits (from surgical and other complications involving the carotid sheath)**

1. ***Hoarseness from damage to CN X—recurrent laryngeal nerve***
2. ***Trapezius muscle weakness from injury to CN XI***
3. ***Tongue weakness from injury to CN XII***
4. ***Inability to swallow—damage to CN XI and CN XII***

Superior Vena Cava

I. GENERAL (201)—The superior vena cava (SVC) receives blood from the head, neck, upper limbs, and thoracic wall. It is located in the superior and middle mediastina.

A. Course—begins behind the right edge of the sternum (from the level of the 1st costal cartilage)—lies anterolateral to the trachea and the hilus of the right upper part of the lung—enters the right atrium (at the level of the 2nd right intercostal space)

B. Tributaries

1. *Right brachiocephalic*
2. *Left brachiocephalic*
3. *Azygos—receives blood from the wall of the thorax*

II. RIGHT BRACHIOCEPHALIC (226)—shorter than the left

A. Course—extends inferiorly behind the right sternoclavicular

B. Tributaries

1. *Internal jugular—the largest vein in the neck—a continuation of the sigmoid sinus. It drains the brain and superficial parts of the face and neck and is located in the carotid sheath.*
2. *Subclavian—a continuation of the axillary*
   i. external jugular—receives blood from the skin of the posterior scalp and neck. It is formed behind the mandible from the retromandibular and posterior auricular.
3. *Inferior thyroid*
4. *Vertebral*
5. *Deep cervical*
6. *Internal thoracic*
7. ***Right lymphatic duct****—enters the brachiocephalic vein near its formation (from the internal jugular and subclavian)*

III. LEFT BRACHIOCEPHALIC (64, 226)

A. Course—runs behind the manubrium to join the right brachiocephalic and form the SVC behind the 1st costal cartilage

B. Tributaries—Internal jugular, Subclavian, Inferior thyroid, Vertebral, Deep cervical, Internal thoracic, Superior intercostal, Esophageal, Tracheal, Superior phrenic, and the **Thoracic duct** which enters the brachiocephalic vein near its formation. The SVC may receive the jugular lymph trunk which receives lymph from the cervical nodes.

Skull

The skull is the skeleton of the head and face. (There are various definitions of the bones constituting the skull, cranium, calvarium, and facial skeleton.)

I. BONES (1–3)

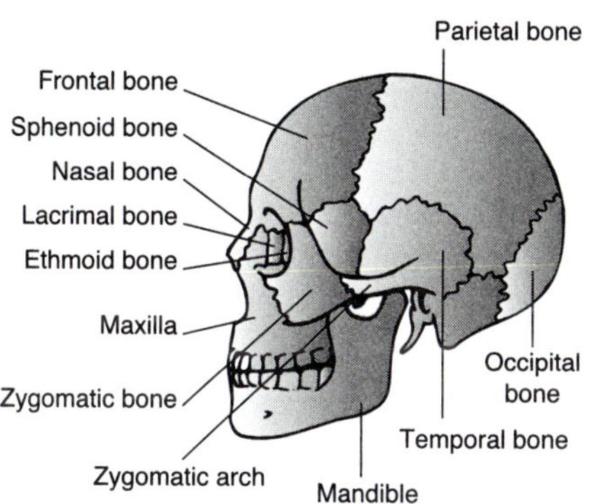

A. Frontal—forms the forehead and the superciliary arch. The orbital plates form the roof of the orbital cavities.

1. *Contains the supraorbital foramen—for the nerve and artery*
2. *Contains the frontal sinuses*

B. Occipital—forms the back of the skull and part of the posterior cranial fossa

1. *Condyles—articulate with the superior facets of the atlas*
2. *The foramen magnum—for the cervical spinal nerves of CN XI, the medulla, and spinal arteries*

C. Parietal—forms the top and sides of the cranium

1. *Contains a groove for the middle meningeal vessels*

D. Temporal (fused from several bones)—forms the sides of the cranium

1. *Contains the mastoid sinus*
2. *Styloid process—for the insertion of neck muscles*
3. *Zygomatic process—articulates with the zygomatic bone*
4. *Petrous portion—on the floor of the middle cranial fossa*
   i. contains the osseous labyrinth
   ii. internal acoustic meatus—transmits CN VII and CN VIII
   iii. between the petrous portion of the temporal bone and the occipital bone are:
      a. foramen lacerum for the internal carotid artery
      b. jugular foramen for the ascending pharyngeal artery, internal jugular vein, and CN IX, CN X, and CN XI

E. Ethmoid—located between the eyes—forms part of the nasal roof

1. *Plates*
   i. perpendicular plate—forms part of the nasal septum (crista galli)
   ii. horizontal (cribriform) plate—forms the roof of the nasal cavities
   iii. lateral plates—form the medial wall of the orbit

2. *Ethmoid sinuses*
    i. anterior middle and posterior
    ii. located lateral to the cribriform plate

F. Sphenoid—separates the cranial cavity from the orbit and anterior cavities

1. *Body—the central part—contains the sphenoid sinus*
    i. greater wings—meet parietal, temporal, and frontal bones at the pterion—form the upper body of the sphenoid
    ii. lesser wings—form the lower body of the sphenoid and the anterior clinoid process (the attachment for the tentorium)
        a. contain the optic canal (foramen) for CN II
    iii. Between the greater and lesser wings of the sphenoid is the superior orbital fissure which transmits CN III, CN IV, CN V (Ophthalmic division), and CN VI.
2. *Pterygoid processes—downward projections from the body—located at the junction of body and greater wings—have medial and lateral plates. The pterygopalatine process is located between the pterygoid and the palatine portions of the maxilla.*
3. *Hypophyseal fossa (Sella turcica)—for the pituitary gland*
    i. The posterior lateral ends of the fossa are called the posterior clinoid processes.

G. Maxillary—forms the front of the face and upper jaw, the lower lateral portion of the nasal cavity, and the floor of the orbit

1. *The anterior nasal spine is the midline meeting of the maxillary bones.*
2. *Maxillary processes*
    i. frontal—an upward projection—joins the frontal bone
    ii. alveolar—holds the upper teeth—forms the alveolar arch
    iii. palatine—forms part of the hard palate
3. *Contains the maxillary sinus (antrum of Highmore). Its anterior wall is thin.*
4. *Contains the infraorbital foramen for the passage of the infraorbital nerve and vessels.*

H. Zygomatic (Zygomic) (Malar)—forms the cheek, the midportion of the zygomatic arch **(a common fracture site)**, and a portion of the outer wall of the orbit—articulates with the greater wings of the sphenoid

I. Nasal—forms the bridge of the nose—articulates with the frontal, ethmoid, and maxillary bones

J. Inferior Nasal Conchae—projects into the nasal cavity

K. Vomer—located in the back of the nasal septum

L. Lacrimal—located in the anteromedial wall of the orbit

M. Palatine—forms part of the orbit, hard palate, and nasal cavity

1. *The greater and lesser palatine foramen for the greater and lesser palatine vessels and nerves*

N. Mandible

1. *Has a body and upper part (alveolar) for the teeth*
2. *Condyles—articulate with the temporal bone (TM joint)*
3. *Has foramen for nerves and arteries to the lower teeth*

➢ II. CRANIUM (4, 5)—the bones of the head

A. Calvaria—superior part—the four upper bones

1. *Sutures*
   - i. coronal—between parietal bones and the frontal bone
   - ii. lambdoid—between the occipital and parietals
   - iii. sagittal—at the top of the skull—between the parietals
2. *Pterion—the area where the frontal, parietal, temporal, and greater wing of the sphenoid come together*
3. *Lambda—point of fusion of the posterior fontanelle (occurs in infancy)—junction of sagittal and lambdoid sutures*
4. *Bregma—point of fusion of the anterior fontanelle (occurs in infancy) at the junction of the coronal and sagittal sutures*
5. *Inion—at the midline apex of the external occipital protuberance*

B. Cranial Base—inferior part—forms the base of the cranial cavity

C. Facial Skeleton

1. *Maxillae*
2. *Zygomatic*

3. *Orbits—composed of 7 different bones*

4. *Nasal cavities—composed of 7 different bones*

D. Mandible

III. FOSSAE (6, 7)

A. Anterior Fossa—formed by the orbital plates of the frontal bone, the cribriform plate, and the jugum of the sphenoid

1. *Contains frontal portions of cerebral hemispheres*

2. *Opening—cribriform transmits the Olfactory nerve (CN I)*

B. Middle Fossa—formed by the body and greater wings of the sphenoid and parts of the temporal and parietal bones

1. *Contains temporal portions of cerebral hemispheres*

2. *Openings*

i. optic canal—for CN II (Optic) and ophthalmic artery

ii. superior orbital fissure—for CN III, CN IV, CN V (Ophthalmic division), CN VI, and superior ophthalmic vein

iii. foramen rotundum—for CN V (Maxillary division) and the accessory meningeal artery

iv. foramen ovale—for CN V (Mandibular division) and the accessory meningeal artery

v. foramen lacerum—for the internal carotid artery which continues on in the carotid canal (along the petrous portion of the temporal bone)

vi. foramen spinosum—for the middle meningeal artery and vein and the meningeal branch of the Mandibular nerve

C. Posterior Fossa—formed by the occipital bone, part of the petrous, a portion of the temporal bone, and the sphenoid portion of the clivus

1. *Contains the cerebellum, pons, and medulla oblongata*

2. *Openings*

i. foramen magnum—for the medulla oblongata, meningeal branches of vertebral arteries, meninges, and the spinal roots of the Accessory nerve (CN XI) (which enters via the foramen magnum but leaves via the jugular foramen)

ii. jugular foramen—for CN IX, CN X, CN XI, and the internal jugular veins

iii. mastoid foramen—for the meningeal branch of the occipital artery and emissary veins

iv. internal acoustic meatus—for CN VII, CN VIII, and the labyrinthine artery. (The Facial nerve traverses the facial canal which ends at the stylomastoid foramen.)

v. hypoglossal canal—for CN XII and the meningeal branch of the ascending pharyngeal artery

IV. CAVITIES (32–34, 38, 39)

A. Cranial Cavity

1. *Middle ear—contains the auditory ossicles (malleus, incus, and stapes) and the proximal end of the eustachian tube*

2. *Internal ear—contains semicircular ducts and the cochlea*

3. *Paranasal sinuses—named according to the bones in which they are found (frontal, maxillary, ethmoid, and sphenoid)*

B. Nasal Cavity—communicates with the paranasal sinuses

1. *Walls of nasal cavities*

i. roof—cribriform plate of the ethmoid

a. transmits branches of CN I

ii. floor—palatine process of the maxillary and horizontal plate of the palatine

iii. medial wall—nasal septum

iv. lateral wall—Three conchae of the ethmoid bone project into the nasal cavity.

2. *Anterior aperture*

i. superior—nasal bones

ii. lateral—frontal processes of the maxilla

iii. medial—septum divides nasal cavity

a. posterior superior part—perpendicular plate of the ethmoid

b. anterior inferior part—cartilage

3. *Posterior aperture*

i. floor—palatine bone

ii. septum—vomer

iii. sides—medial pterygoid plate of the sphenoid bone

C. Orbital Cavity

1. *Walls*

i. roof—orbital plate of the frontal bone

ii. floor—orbital plate of the maxillary bone

iii. medial—maxillary, lacrimal, ethmoid, and palatine bones

iv. lateral—zygomatic bone (frontal process)

v. posterolateral—sphenoid

2. *Orbital fissures*

i. superior orbital fissure

ii. inferior orbital fissure—located between the floor and lateral wall of the orbit and opens into the infratemporal fossa behind the maxilla

3. *Foramen*

i. optic—leads into the middle cranial fossa

ii. ethmoidal (2)—lead into the anterior cranial fossa

➢ V. SCALP MUSCLES (21)

A. Epicranius (Occipitofrontalis)

1. *Occipital—extends from the superior nuchal line to the skin—moves the scalp backward*

2. *Frontal—from the frontal bone near the coronal suture to the skin—moves the scalp backward*

B. Auriculares (Anterior, Middle, and Posterior)—extend from areas anterior, superior, and posterior to the ear to go to the auricle

➢ VI. MUSCLES WHICH MOVE THE HEAD (161)

A. Head Flexors—Sternocleidomastoids—from the clavicles and sternum to the mastoid processes. Either of these two muscles working alone moves the head to the side; both muscles working together flex the head on the chest.

B. Head Extensors—Either of these two muscles working alone moves the head to the side; both muscles working together extend the head.

1. *Longissimus—from the lower cervical and upper thoracic vertebrae to the mastoid process*

2. *Splenius—from the lower cervical and upper thoracic vertebrae to the mastoid process*

3. *Semispinalis—from the lower cervical and upper thoracic vertebrae to the occipital bone*

Brain

I. **PARTS (99–101, 105–109, 140)—The medulla, pons, and cerebellum form the hindbrain. The forebrain consists of the cerebral hemispheres and the diencephalon. The brain stem consists of the medulla, pons, mesencephalon, and diencephalon.**

A. Medulla Oblongata (108–109)—between pons and spinal cord—located in the posterior cranial fossa—contains the reticular formation which is involved in the control of breathing, cardiac rate, respiration, and digestion

1. *Centers*

   i. nuclei cuneatus and gracilis—for sensory impulses from the spinal cord (pain and temperature)

   ii. for CN IX to CN XII—regulatory (respiration, swallowing, vomiting, and cardiovascular)

   iii. solitarius nucleus—for taste

2. *Features*

   i. inferior cerebellar peduncle—at the upper posterior part of the medulla—connects medulla and cerebellum

   ii. crossing of the corticospinal tract (decussation of the pyramids). Fibers connecting the spinal cord with higher levels cross over in the medulla.

   iii. cavity—part of the 4th ventricle

B. Pons (108, 111)—located between the medulla and cerebral peduncles

1. *Centers—for CN V to CN VIII—pontine nuclei relay information to the cerebellum from the cerebral cortex*

   i. facial motor nucleus

   ii. superior salivary nucleus

   iii. vestibular and cochlear nuclei for balance and hearing

2. *Features*

   i. The superior cerebellar peduncle is above the 4th ventricle.

   ii. The pons connects the right and left halves of the cerebellum and connects each side of the cerebellum with the opposite cerebral hemisphere.

   iii. involved in REM sleep and arousal

C. Mesencephalon (Midbrain) (101–103)—connects the cerebrum with the opposite cerebellar hemisphere (in the posterior cranial fossa)

1. *Centers—for CN III, sensory portions and proprioception of CN V. The midbrain contains the red nucleus and substantia nigra.*
2. *Features*
   i. colliculi—reflexes for hearing and sight
   ii. cerebral peduncles—motor fibers to the spinal cord
   iii. cavity—cerebral aqueduct (of Sylvius) connects the third and fourth ventricle
   iv. corpora quadrigemina—consists of 2 superior and 2 inferior colliculi

➢ D. Diencephalon (105, 106, 140)—consists of the thalamus, hypothalamus, epithalamus, and subthalamus—lies lateral to the 3rd ventricle

1. *Centers*
   ➢ i. thalamus (105)—center for optic tract fibers. The medial geniculate nucleus is for auditory pathways.
   ➢ ii. hypothalamus (106, 140)—for metabolic regulation (it influences the pituitary) and autonomic nervous system regulation (including regulation of blood pressure and body temperature)—has tracts to the thalamus, brain stem, spinal cord, and pituitary
      a. amygdala—regulates heartbeat and some visceral functions and processes the emotion of fear—associated with emotional memory and reward
      b. hippocampus—aids in the establishment of long-term memory in regions of the cerebral cortex
      c. paraventricular and supraoptic nuclei—peptinergic neurons synthesize oxytocin and vasopressin
2. *Features*
   i. thalamus—a relay station for sensory and motor signals coming to and going fròm the cerebral cortex
   ii. pineal body—attached to the roof of the 3rd ventricle
   iii. mammillary body—for smell, recognition, and recall
   iv. hypothalamicohypophysial tract

➢ E. Hypophysis (140)—lies in the sella turcica, near the optic chiasma

1. *Anterior lobe (Adenohypophysis)—produces trophic and stimulating hormones. It is derived from ectoderm.*

i. pars distalis—also receives blood supply from the hypophysial portal system

ii. pars infundibulum—attaches the hypophysis to the hypothalamus. It is surrounded by the pars tuberalis.

iii. pars intermedia

2. *Posterior lobe (Neurohypophysis)—releases oxytocin and vasopressin. It is derived from neuroectoderm.*

i. pars nervosa—receives hypothalamic nerve fibers

ii. infundibulum—contains pituicytes

F. Cerebellum (109)—Located in the posterior cranial fossa, it consists of lateral lobes and a central mass and is covered by occipital lobes of the cerebral hemispheres. It is essential for movement and coordination. It has 2 lobes connected by the vermis.

1. *Centers—fastigial, globose, emboliform, and dentate nuclei*

2. *Features*

i. arbor vitae—white matter

ii. peduncles (superior, middle, and inferior)—connections to brain stem (mesencephalon, pons, and medulla)

iii. tracts for touch, temperature, and proprioception

G. Cerebrum (99–101)

1. *Centers*

i. Hemispheres—right and left

a. lobes

1) frontal—located in front of the central sulcus—involves planning, language expression, and speech (in Broca's speech area). The basal forebrain influences arousal and emotions.

2) parietal lobe—located between the postcentral sulcus and the occipital lobe (i.e., behind the central sulcus and above the posterior portion of the lateral sulcus)—involves touch, taste, and proprioception

3) occipital—located posterior to the parietal and temporal lobes—a primary visual area

4) temporal—located below the posterior ramus of the lateral sulcus—contains Wernicke's area for understanding language—analyzes sound and is

involved with taste and smell—contains the lateral geniculate body for the sense of sight

2. *Features*

   i. corpus callosum—a large band of nerve fibers through which information flows back and forth between the right and left hemispheres—connects the hemispheres

II. VENTRICLES (102, 103, 109)—communicate with each other and with the subarachnoid space—contain the choroid plexus which elaborates cerebrospinal fluid

A. Lateral Ventricles—separated by the septum pellucidum

1. *Horns*

   i. anterior horn—into frontal lobe

   ii. posterior horn—into occipital lobe

   iii. inferior horn—into temporal lobe

2. *Each lateral ventricle connects with the 3rd ventricle via the* ***Interventricular Foramen of Monro***

B. Third Ventricle—located in the diencephalon—connects with the 4th ventricle via the **cerebral aqueduct of Sylvius**

C. Fourth Ventricle—located between the cerebellum and pons. The floor of the 4th ventricle is formed by the medulla. It connects with the arachnoid space via 2 lateral apertures **(foramen of Luschka)** and 1 median aperture **(foramen of Magendie)** all of which drain into the **cisterna magna.**

III. FISSURES (SULCI) (99, 101)

A. Longitudinal Cerebral—between hemispheres—contains the falx

B. Central Sulcus (of Rolando)—arises at the upper margin of the longitudinal cerebral fissure and runs anterior, inferior, and toward the lateral sulcus—separates motor and sensory areas

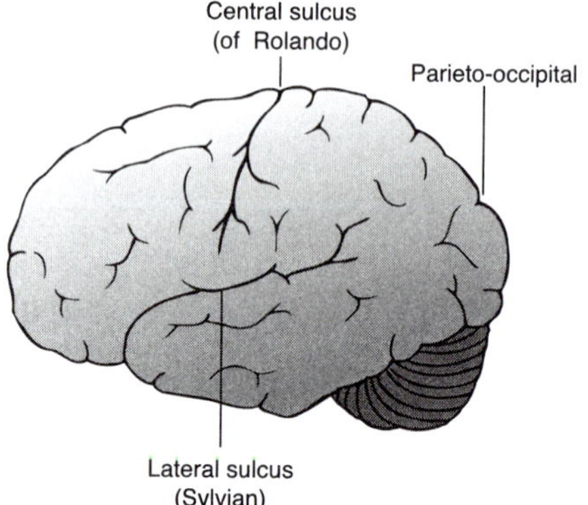

C. Lateral Sulcus (Sylvian)—separates frontal and temporal lobes and approximately divides the cerebrum into top and bottom areas. It is opposite the pterion and has 3 branches.

D. Parieto-occipital—separates parietal and occipital lobes—located above the lambda in front of the occipital lobe

E. Temporal Sulci—divide the temporal lobe into 3 gyri (superior, middle, and inferior)

F. Calcarine Sulcus—on medial aspect of cerebral hemispheres and within the occipital lobe

G. Transverse Cerebral—separates cerebral hemispheres from the cerebellum, midbrain, and diencephalon—contains the tentorium

IV. COVERINGS (96)

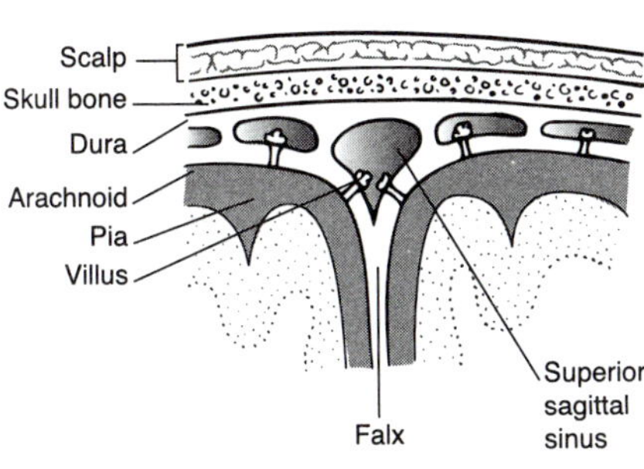

A. Pia—next to the cortex

1. *A thin layer transparent vascular area which follows sulci and gyri. It folds into the ventricles as the* ***choroid plexus****.*

2. *Subarachnoid space—between the pia and arachnoid. In some areas the subarachnoid space is larger. These areas are called cisterns. A large one located between the cerebellum and medulla is called the* ***cisterna magna****.*

B. Arachnoid—lies between pia and dura—does not enter sulci

1. *Arachnoid villi extend thru the walls of the dural venous sinuses. Most protrude into the superior sagittal sinus.*

C. Dura—outermost covering—lines the skull inner surface—consists of two layers (the outer is attached to bone)

1. *Subdural space—between arachnoid and dura*

2. *Forms septa*

i. falx cerebri—located between cerebral hemispheres—from the crista galli of the ethmoid bone to internal occipital protuberance posteriorly—encloses the superior and inferior sagittal sinus

ii. tentorium cerebelli—located between the occipital lobes and cerebellum—extends from the temporal bone and clinoid process of the sphenoid bone to the occipital bone—encloses the transverse sinuses

iii. falx cerebelli—located between cerebellar hemispheres

V. ARTERIES TO THE BRAIN (130, 131, 134, 136)—Branches of the internal carotids supply the anterior portion of the brain. Branches of the vertebrals supply the brain stem, cerebellum, and portions of the temporal and occipital lobes.

A. Internal Carotid—enters middle cranial fossa via carotid canal and gives off the following cerebral branches:

1. *Anterior cerebral—to anterior-middle (midline) portion of the cerebral hemisphere*

i. anterior communicating—connects the anterior cerebrals

It is subject to aneurysms.

2. *Middle cerebral—to lateral surface of the cerebral hemispheres*

3. *Posterior communicating—joins the posterior cerebral of the basilar*

B. Vertebral—The right and left vertebral arteries join to form the **basilar** artery which terminates by splitting into the right and left superior cerebellar and posterior cerebral arteries. These arteries supply the superior portions of the cerebellum and the occipital and medial portions of the temporal lobes, respectively.

VI. ARTERIES TO THE MENINGES (95)

A. Anterior Meningeal—from the ethmoidal of the internal carotid—supplies the anterior fossa dura

B. Middle Meningeal—a branch of the maxillary artery—enters cavity via foramen spinosum (of the sphenoid bone)

1. *Anterior branch to frontal lobe*
2. *Posterior branch to parietal lobe—extends horizontally interior to the temporal and parietal bones*

Bone trauma may result in an epidural hematoma. Continued hematoma expansion here is characterized by a decreasing level of consciousness.

C. Posterior Meningeal—a branch of the occipital. It enters thru the jugular foramen and supplies the posterior fossa dura.

VII. VENOUS (DURAL) SINUSES (97, 98, 137)—between layers of dura—no valves. Flow in sinuses is decreased by trabeculae.

A. Sagittal Sinuses—extend along the falx

1. *Superior sagittal sinus. Blood passes backward in the falx to join the straight sinus and the right transverse sinus (at the confluence of sinuses); thence blood enters the sigmoid sinus and then the internal jugular vein.*
2. *Inferior sagittal sinus—joins the great cerebral vein (of Galen) to become the left transverse sinus which enters the sigmoid sinus and then the internal jugular vein*

B. Transverse Sinuses—receive veins from the bones (diploic veins), cerebellum, and cerebrum. The right transverse sinus receives mostly superior sagittal sinus blood while the left receives mostly straight sinus blood.

C. Sigmoid Sinuses—the distal portions of the transverse sinuses. A sigmoid sinus enters (continues on as) the internal jugular vein.

D. Straight Sinus—receives blood from the Great Cerebral Vein and inferior sagittal sinus

E. Cavernous Sinuses—located around the sella turcica, on the sides of the sphenoid bone. Each sinus extends from the superior orbital fissure to the petrous portion of the temporal bone. Each sinus connects with the ophthalmic veins via emissary and facial veins. **These connections may allow facial infections, especially in the midline, to enter the brain.**
The cavernous sinuses drain into the transverse sinuses.

F. Occipital Sinus—located behind the foramen magnum

G. Basilar Sinus—located in front of the foramen magnum

H. Petrosal Sinuses

1. *Superior—drains into a transverse sinus*

2. *Inferior—drains into a sigmoid sinus*

➢ **VIII. CONFLUENCE OF SINUSES** (97)**—formed by the straight sinus, superior sagittal sinus, occipital sinus, and transverse sinuses**

➢ **IX. VEINS** (94, 137–139)**—do not accompany arteries. They have thin walls, no muscular coats, and no valves.**

A. Cerebral Veins—**Rupture of the cerebral veins may cause a subdural hematoma.**

1. *Superior cerebral—drains the hemispheres and enters the superior sagittal sinus*

2. *Middle (Superficial) cerebral—drains the lateral surfaces of the hemispheres and enters the cavernous sinus*

3. *Inferior cerebral—drains inferior portions of hemispheres*

4. *Great cerebral vein (of Galen)—receives deep cerebral veins and drains into the straight sinus*

B. Cerebellar Veins—drain into posterior venous sinuses

C. Meningeal Veins—located between the dura and the skull bone to empty into the dural sinuses

1. *Middle meningeal—via the foramen spinosum and foramen ovale to enter the pterygoid plexus*

2. *Superior and Inferior petrosal*

D. Diploic Veins—located between the interior and exterior tables of the bones of the calvarium. They drain into dural sinuses and emissary veins. (Emissary veins connect the dural sinuses with veins outside of the calvaria. **This allows a path for infection.**)

TABLE 2-1. CRANIAL NERVES

NERVE	FROM	OPENING IN SKULL	TO
CN I **Olfactory** (113) (Sensory)	Olfactory area in nasal cavity	Cribriform plate of the ethmoid	Olfactory bulb to cerebrum (corpus callosum and uncus)
CN II **Optic** (114) (Sensory)	Retina of eye	Optic canal	Optic chiasma to optic tracts to the brain stem to the occipital lobe
CN III **Oculomotor** (115) (Motor)	Midbrain	Superior orbital fissure	Eyeball muscles (except 2) and smooth muscle of iris and ciliary body via the ciliary ganglion
Parasympathetic	Midbrain	Superior orbital fissure	Smooth muscle of iris and ciliary body
CN IV **Trochlear** (115) (Motor)	Midbrain	Superior orbital fissure	Superior oblique of the eye
CN V **Trigeminal** (116)			
Ophthalmic (40) (Sensory)	Orbit, frontal sinus, skin of upper eyelid, nose, and anterior ½ of scalp	Superior orbital fissure	Brain stem
Maxillary (40) (Sensory)	Eyelid, nose, cheek, jaw	Foramen rotundum	Brain stem
Mandibular (41) (Sensory)	Lower lip, jaw, tongue, face	Foramen ovale	Brain stem
Mandibular (41) (Motor)	Brain stem—pons	Foramen ovale	Muscles of mastication: 2 neck muscles 3 pharyngeal muscles
CN VI **Abducent** (115) (Motor)	Brain stem—pons	Superior orbital fissure	Lateral rectus of the eyeball
CN VII **Facial** (117) (Motor)	Brain stem—pons	Stylomastoid foramen	Muscles of expression, posterior digastric, stylohyoid, stapedius
Parasympathetic	Pons	Internal auditory meatus	Submandibular, sublingual, lacrimal glands
(Sensory)	Taste buds of anterior ⅔ of tongue, soft palate	Internal auditory meatus	Medulla
CN VIII **Vestibulocochlear** (118) (Sensory)	Vestibular—Semicircular canals Acoustic—Cochlear duct and spiral organ	Internal auditory meatus (located in the petrous part of temporal bone)	Area between medulla and pons
CN IX **Glossopharyngeal** (119) (Motor)	Medulla	Jugular foramen	Stylopharyngeus and superior constrictor of the pharynx

TABLE 2-1. *Continued*

NERVE	FROM	OPENING IN SKULL	TO
(Sensory)	Mucosa of pharynx, back of tongue, epiglottis, soft palate, tonsils	Jugular foramen	Medulla
	Carotid sinus and carotid body	Jugular foramen	Medulla
Parasympathetic	Medulla	Jugular foramen	Parotid gland via the otic ganglion
CN X **Vagus** (120)	Medulla	Jugular foramen	Pharyngeal plexus to muscles of larynx and pharynx
Parasympathetic			Thoracic and abdominal viscera
CN XI **Accessory** (121) (Motor)	Medulla	Jugular foramen	Joins Vagus then to pharyngeal plexus
	Spinal (C1–C5)	Ascending thru the foramen magnum	Sternocleidomastoid and trapezius via jugular foramen
CN XII **Hypoglossal** (122) (Motor)	Medulla	Hypoglossal canal (Anterior condylar canal)	Tongue muscles, sternohyoid, sternothyroid, omohyoid

TABLE 2-2. CRANIAL NERVE INJURIES

NERVE AND POSSIBLE INJURY SITE	IMPAIRMENT
CN I. Olfactory gyrus or cribriform plate	Decreased sense of smell
CN II. Lesion of the right optic tract	Blind halves of the right sides of both retinas → blind halves of both lateral visual fields
Lesion of the fibers crossing at the optic chiasma usually due to a pituitary tumor	Loss of vision in both lateral visual fields
CN III. Compression of the uncus of the temporal lobe due to herniation	Pupil dilation (mydriasis) on the same side Weakness of the extrinsic ocular muscles Levator palpebrae weakness → eyelid ptosis
CN IV. As it emerges from the dorsal aspect of the brain and decussates completely	Superior oblique muscle → inability to look down and out
CN V. Trigeminal ganglion or nucleus	Tic douloureux (pain in one or more divisions of the Trigeminal nerve)
Ophthalmic in superior orbital fissure	Decreased sensation of skin of forehead
Maxillary in foramen rotundum	Sensory changes over maxilla
Mandibular near the tragus	Weak muscles of mastication
Mandibular below the angle of the mandible	Lip droop
CN VI. Pons-medulla junction	Lateral rectus muscle → lateral gaze paralysis
CN VII. Cerebral hemorrhage affecting the facial motor nucleus	Weakness of contralateral facial muscles
	Loss of the efferent limb of the blink reflex on the same side
Facial nerve	Bell's palsy (weakness of muscles of facial expression on same side)
CN VIII. Lateral lemniscus	Decreased hearing in both ears
Internal auditory meatus	Decreased hearing in same ear Tinnitus, vertigo
CN IX. Jugular foramen	Loss of carotid reflex, gag reflex, and sensation in the upper pharynx and palate
CN X. Jugular foramen	Tachycardia, vomiting, decreased respiration
Injury on one side	Weakness of the palate levator (therefore the uvula points to the normal side)
Posterior aspect of the carotid sheath (damage from blunt trauma or endarterectomy)	Hoarseness
CN XI. Jugular foramen or neck surgery	Trapezius muscle weakness → inability to shrug the shoulders
CN XII. Condylar canal or as the nerve passes over the carotid bulb and lies under the facial vein	Weakness of the tongue muscles on the same side resulting in protrusion to that side

Orbit

I. **FASCIA (78)—located between the orbit and eyeball—extends from the optic nerve to the sclerocorneal junction. An anterior fascial extension forms a sheath for the extrinsic eye muscles.**

II. **BONES (1)—covered with periosteum which lines the orbit**

A. Walls

1. *Roof—orbital plate of the frontal bone*
2. *Floor—orbital plate of the maxillary bone*
3. *Medial—maxillary, lacrimal, ethmoid, and sphenoid bones*
4. *Lateral—zygomatic bone (frontal process) and the greater wing of the sphenoid*

B. Fissures

1. *Superior orbital fissure (1, 7)—located between the roof and lateral wall posteriorly—transmits CN III, CN IV, the Ophthalmic branch of CN V, CN VI, and the superior ophthalmic vein*
2. *Inferior orbital fissure (1, 2)—located between the floor and lateral wall. It opens into infratemporal fossa behind the maxilla.*

C. Foramen

1. *Optic canal (7, 81)—located in the sphenoid bone (lesser wings)—leads into the middle cranial fossa—transmits the Optic nerve (CN II) and the ophthalmic artery*
2. *Infraorbital foramen (1)—located* ***below*** *the orbit in the maxillary bone—transmits the infraorbital nerves and vessels and the maxillary nerve*
3. *Lacrimal foramen—in the sphenoid bone*

D. Notches

1. *Supraorbital notch (foramen) (1)—located on the supraorbital margin—for the supraorbital branches of the ophthalmic artery and nerve*
2. *Trochlear notch—for the superior oblique muscle "pulley"*

Eye

I. EYEBALL (82–84)

A. Outer Layer—fibrous coat

1. *Cornea—anterior transparent portion thru which the pupil and iris can be seen*

   Sensation via Ophthalmic division of the Trigeminal

2. *Sclera—posterior, white and opaque*

B. Middle Layer

1. *Iris—the pigmented diaphragm*

   i. surrounds the pupil and separates the anterior and posterior chambers

   ii. innervated by parasympathetic fibers

   iii. muscles

   a. sphincter—circularly oriented—parasympathetic innervation

   b. dilator—radially oriented—sympathetic innervation

2. *Choroid—the vascular layer between the sclera and retina*

3. *Ciliary body*

   i. via the suspensory ligament, suspends the lens and contains the ciliary muscle which changes the shape of the lens (the ocular chamber is anterior to the lens—the large vitreous chamber is posterior to the lens)

   ii. ciliary process—secretes aqueous humor which is contained in both the anterior and posterior chambers

C. Retinal Layer—innermost layer contains rods and cones—continuous with the optic tract

1. *Optic disk—located in the optic fundus—contains retinal arteries and veins*

2. *Macula lutea—contains the fovea centralis (located lateral to the optic disk) for sharp vision*

3. *Optic fundus—posterior portion of the retinal layer*

II. CONJUNCTIVA (78)—mucous membrane—innervated by branches of the trigeminal nerve

A. Bulbar—covers the sclera—thin

B. Palpebral—vascular—thick—covers the inner eyelid surface

III. MUSCLES (79)

A. Extrinsic—skeletal muscle extends from a fibrous ring around the optic foramen to insert on the sclera

1. *Rectus muscles (4)—superior, inferior, medial (adductors), and lateral rectus (abductor)*
2. *Superior oblique—makes a turn at the trochlea (a fibrous pulley). It rotates the eyeball outward and downward.*
3. *Inferior oblique—rotates the eyeball outward and upward*

B. Intrinsic—smooth muscle

1. *Sphincter and dilator of the pupil*

IV. OPHTHALMIC ARTERY (80)—from the internal carotid

A. Retinal (central artery of the retina)

B. Ciliary—The posterior ciliary artery supplies the middle layer of the eyeball.

C. Muscular—to the eyeball muscles and iris

V. VEINS (80)—Superior and inferior ophthalmic veins drain into the cavernous sinus. They have anastomoses. The superior ophthalmic vein accompanies the ophthalmic artery.

VI. NERVES (81, 115, 126)

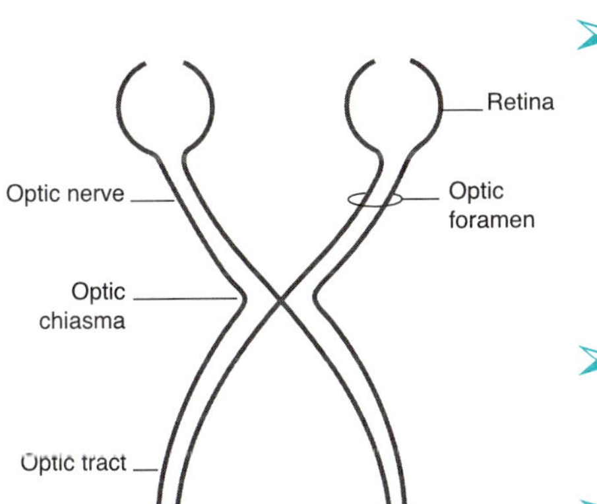

A. ANS (115, 126)

1. *Parasympathetic—via CN III to the ciliary ganglion (located between the optic nerve and lateral rectus muscle) and thence to the ciliary muscle via the short ciliary nerve. The ciliary nerve goes to the cornea and sclera.*
2. *Sympathetic—to the pupil dilator via the long ciliary nerve*

B. Sensory—CN II (Optic nerve) (114)—fibers cross at the optic chiasma (which lies above and anterior to the hypophysis). The dural sheath of the optic nerve is continuous with the sclera.

C. Motor (115)—enter the orbit via the superior orbital fissure

1. *CN VI (Abducent)—to the lateral rectus muscle*
2. *CN IV (Trochlear)—to the superior oblique muscle*
3. *CN III (Oculomotor)—to the remainder of the extrinsic muscles and the superior palpebral elevator muscle. It passes between the superior cerebellar and posterior cerebral arteries. Its cell bodies are located in the oculomotor nucleus.*

VII. MEDIAL CANTHUS (77)

A. Lacrimal Lake—red area of conjunctiva at the medial eye angle

B. Lacrimal Caruncle—small red body filling the lacrimal lake. It contains glands which secrete whitish material.

C. Plica Semilunaris (Semilunar Fold)—a fold of palpebral conjunctiva lateral to the caruncle. The plica overlaps a small area of the eyeball.

D. Lacrimal Sac—the upper dilated end of the nasolacrimal duct

E. Nasolacrimal Duct—drains lacrimal canaliculi into the inferior meatus of the nose

VIII. GLANDS (40, 77)

A. Tarsal—located in the tarsal plate. They lubricate eyelid edges.

B. Ciliary—sebaceous glands of the eyelashes

C. Lacrimal—located in the upper outer portion of the orbit

1. *The course of tears—glands ≫ lake ≫ punctum ≫ sac ≫ nasolacrimal duct ≫ inferior nasal meatus*

2. *Parasympathetic innervation, increases secretions, via CN VII (Facial) and the greater superficial petrosal nerve to the pterygopalatine ganglion and then to the lacrimal gland*

3. *Sympathetic innervation (decreases secretions) via superior cervical sympathetic ganglion and internal carotid nerve to the deep petrosal nerve to the pterygopalatine ganglion and thence to the lacrimal gland*

IX. EYELIDS (76)

A. Palpebral Fissure—opening between eyelids

B. Muscles

1. *Levator palpebrae superioris—raises the upper eyelid and is innervated by CN III (Oculomotor)*

2. *Orbicularis oculi—closes the eyelid—innervated by CN VII,* ***involved in the "corneal blink reflex"***

C. Lymphatics—to the submandibular nodes

D. Tarsus—dense connective tissue in the eyelids

E. Tarsal Glands—located in the tarsal plate between tarsi and conjunctivae. Their ducts open on free margin of the lids.

F. Lacrimal Papilla—small elevation, about 5 mm from the angle of the eye, which contains an opening called the lacrimal punctum

Ear

The "ear" includes external, middle, and internal (inner) ears.

I. EXTERNAL EAR (87, 88)

A. Auricle (Pinna)—elastic cartilage covered with skin

1. *Arteries—posterior auricular (from the external carotid) and superficial temporal*
2. *Veins—auricular to external jugular*
3. *Lymphatics go to the parotid and cervical nodes.*
4. *Nerves*
    i. sensory—the great(er) auricular (from C2 to C3) and the auriculotemporal (from the Mandibular of CN V)
    ii. motor—from branches of the Facial nerve (CN VII)

B. Canal (External Auditory Meatus)—runs medial forward and downward—1 inch long—innervated by CN IX, CN X, and the greater auricular nerve (C2–C3)

1. *Cartilaginous—outer ⅓ contains hair and glands (secretes cerumen)*
2. *Bony (inner ⅔)—located in the temporal bone—related to the TM joint and the parotid gland*

II. MIDDLE EAR (87–89)—located in the petrous part of the temporal bone. It consists mostly of the tympanic cavity which is connected to the nasopharynx by the eustachian (auditory) tube. The eustachian tube, then, extends from the anterior wall of the middle ear to the nasopharynx where it is related to lymphoid tissue. It is also related to the tensor veli palatini muscle.

A. Walls

1. *Lateral—eardrum (tympanic membrane)*
2. *Posterior—opens to mastoid air cells*
3. *Anterior—contains opening for eustachian tube*
4. *Floor—bone next to jugular fossa*
5. *Roof—thin bone next to middle cranial fossa*

B. Windows—openings in the vestibule of the osseous (bony) labyrinth

1. *Oval window—connected to foot piece of the stapes*
2. *Round window—located beneath the oval window*

C. Cavity—contains bones which transmit sound waves

1. *Malleus (hammer)—articulates with the incus*

   i. attached to the tympanic membrane by the tensor tympani muscle which is innervated by the Trigeminal nerve

2. *Incus (anvil)—connects the malleus and stapes*

3. *Stapes (stirrup)*

   i. The stapedius muscle, from the stapes to the temporal bone, is innervated by the facial nerve.

   ii. The stapes is attached to the oval window by a ligament.

D. Branches of the External Carotid to the Middle Ear

1. *Tympanic branch of maxillary*

2. *Stylomastoid of posterior auricular*

3. *Petrosal of middle meningeal*

4. *A branch of the ascending pharyngeal*

III. INNER EAR (87, 90–92)—a sensory apparatus consisting of 2 labyrinths located within the petrous portion of the temporal bone. The inner ear contains:

A. Osseous Labyrinth—perilymph is located between the osseous labyrinth and membranes

1. *Cochlea contains:*

   i. scala tympani—ends at the round window

   ii. scala vestibuli (vestibular duct)—contains perilymph—begins at the oval window

2. *Vestibule—communicates with the semicircular canals and cochlea. It contains the utricle and saccule and also contains the oval window.*

3. *Semicircular canals (3)—located at right angles to one another. Their ends open into the vestibule.*

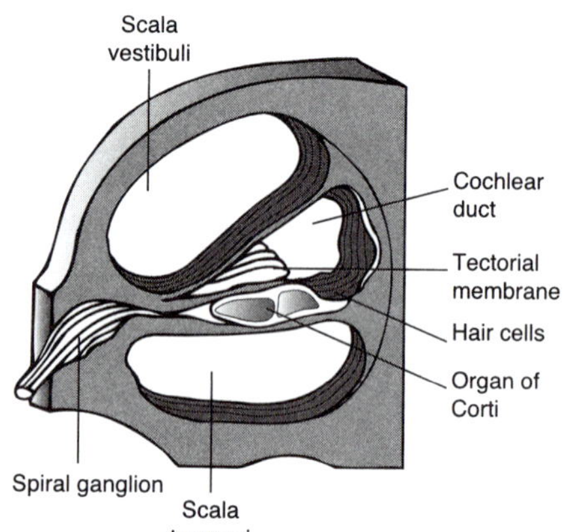

B. Membranous Labyrinth—located in the bony (osseous) labyrinth. The membranous labyrinth is a sac which contains endolymph. It is concerned with the sense of balance.

1. *Vestibular portion—contains maculae.*
   The maculae contain sensory hair cells influenced by positioning (gravity)—i.e., head movements displace (move) the vestibular sensory hair cells.

i. utricle—The semicircular ducts unite here.

ii. saccule—The end of the saccule is called the **endolymphatic duct** (it lies between the utricle and saccule).

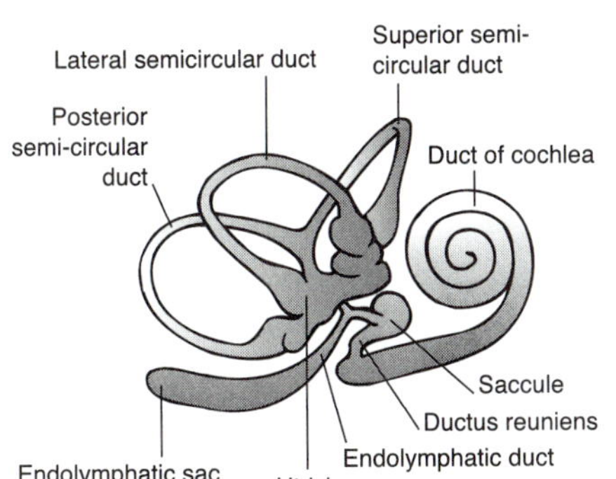

2. *Semicircular ducts (3)—in the semicircular canals*

   i. one in each bony canal

   ii. Their dilated ends are connected to the utricle and are called **ampullae** (cristae ampullaris).
   The ampullae respond to rotational movements.

3. *Cochlear portion*

   i. the cochlear duct (scala media)

      a. connects the cochlea to the saccule

      b. Its epithelium contains cochlear hair cells and is called the **Organ of Corti.**

      c. contains endolymph and is surrounded by perilymph

   ii. The **ductus reuniens** (lower end of the cochlear duct) unites the vestibular and cochlear portions of the membranous labyrinth and connects the saccule with the cochlear duct proper.

C. Arteries to the Inner Ear:

1. *Internal auditory (Labyrinthine)—from the basilar*

2. *Stylomastoid—from the posterior auricular from the external carotid*

D. Veins of the Labyrinth—drain into the internal auditory veins which may then drain into a petrosal sinus or the transverse sinus or the sigmoid sinus

IV. NERVES TO THE LABYRINTH—VESTIBULOCOCHLEAR (CN VIII) (87, 90, 91)

A. Vestibular Divisions

1. *Nerve of equilibration*

2. *From the utricle, saccule, and ampullae of the semicircular ducts to the vestibular ganglion*

B. Cochlear Division—Impulses go from hair cells of Organ of Corti (spiral organ) to cells of the spiral ganglion.

C. A. and B. merge to form CN VIII which passes thru the petrous portion of the temporal bone to reach the brain stem.

Face

The face is the area between hairline, ears, and chin.

I. NOSE (31)—a pyramidal appendage—has a root, tip, and wings (alae)

A. Lateral Walls—alae

B. Base—2 openings (nares) with a septum (columna)

C. Bridge—supported by nasal bones

D. Ophthalmic vessels and nerves supply the nose.

II. BONES (31, 33)—maxilla, palatine, mandible, and zygomatic

III. MUSCLES OF EXPRESSION (20, 21)—located around the eyes, ears, nose, and mouth—innervated by the Facial nerve

A. Orbicularis Oris—around the mouth—closes the lip

B. Orbicularis Oculi—around the eye

1. *Orbital portion—across the forehead, temple, and cheek*
2. *Palpebral portion—closes the lids*
3. *Lacrimal portion—keeps the lids attached to the eyeballs and draws the eyebrows down*

C. "Smiling" Muscles—upper lip elevators

1. *Levator labii superioris—from the maxilla to the upper lip*
2. *Zygomaticus (major and minor)—from the zygoma to the lip corners*

D. "Grinning" Muscles

1. *Depressor labii inferioris—from the mandible to the orbicular oris muscle*
2. *Levator anguli oris—from the inferior portion of the orbit to the angles of the mouth*
3. *Risorius—from fascia over masseter to the angle of the mouth*

E. "Wrinkle Muscles"—Procerus and Corrugators

F. Mentalis—chin muscles—from the mandible to the chin—protrude the lower lip

G. Buccinator—from the jaws near the lower teeth to the lips (lies deep to risorius and the zygomaticus major)—aids in mastication but is supplied by the **Facial** nerve, and therefore is a muscle of expression

IV. EXTERNAL CAROTID BRANCHES TO THE FACE (17, 31)

A. Facial—from the external carotid

1. *Superior labial—to the upper lip and side of alae*
2. *Inferior labial—to the lower lip*

B. Superficial Temporal—begins between the parotid and the back of the mandible to end in the scalp

1. *Transverse facial—to the parotid and skin*
2. *Terminal branches to upper superficial muscles and skin*

V. VEINS (17)

A. Anterior Facial (angular)—formed near the inner angle of the eye. The anterior facial vein drains into:

1. *Diploic veins—via frontal veins*
2. *Pterygoid plexus—via deep facial veins—drains into maxillary vein and cavernous sinus*
3. *Cavernous sinus—via superior ophthalmic veins*

Midline infections of the face may extend into the cavernous sinus. It is also subject to thrombosis.

VI. LYMPHATICS—above the eye may drain into the parotid nodes—below the eye may drain into the submandibular nodes. Lymphatics of the cheeks drain into parotid and submandibular nodes. Lymphatics of the lower lip drain into submental nodes.

VII. LYMPH NODES (66)—Individual nodes which may subsequently send lymphatics to the parotid and submandibular nodes include nodes located in the front of the tragus, below the orbit, on the buccinator muscle, and on the outer surface of the mandible.

A. Anterior Auricular—in front of the tragus—drains the pinna and temporal area—has efferents to parotid and deep cervical nodes

B. Facial—located below the orbit, on the buccinator muscle and on the outer surface of the mandible—drains the eyelids, cheek, nasal skin and mucosa and has efferents to submandibular nodes

C. Parotid—located in the parotid gland—drains the nose, eyelids, forehead, and tympanic cavity—has efferents to deep cervical nodes

D. Submandibular—below the mandible—drains the infraorbital and chin area—has efferents to deep cervical nodes

E. Submental—located on the digastric muscle. They drain the lower lip and the floor of the mouth. Submental efferents go to submandibular and cervical nodes.

VIII. NERVES (41, 65)

A. Sensory—branches of the Trigeminal nerve (CN V) to the face

1. *Ophthalmic—from the top of the head to the tip of the nose*
   i. nasociliary—to tip and root of nose
   ii. infratrochlear—to skin around the eyelid and nose
   iii. frontal
      a. supratrochlear—to the medial part of upper eyelid and forehead
      b. supraorbital—leaves the orbit thru the supraorbital notch, to supply the central portion of the upper eyelid, skin of the forehead, and the scalp to the vertex
   iv. lacrimal—to the lateral upper eyelid and lacrimal gland

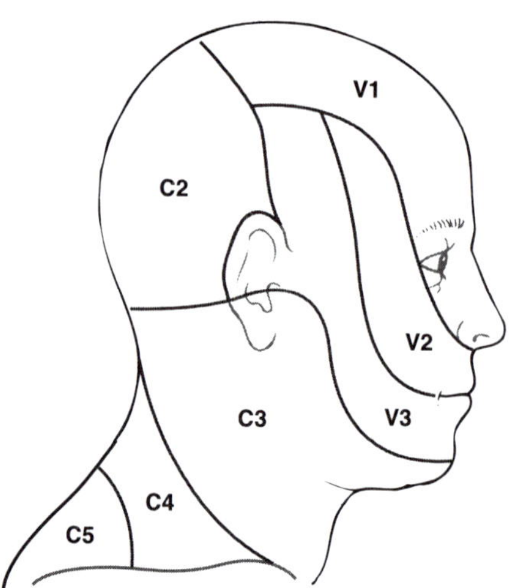

2. *Maxillary—extends from the semilunar ganglion thru the foramen rotundum over the pterygopalatine fossa thru the infraorbital fissure along the floor of the orbit thru the infraorbital foramen (generally sensory over the maxilla)*
   i. infraorbital—to lateral nose, upper lip, and lower eyelid
   ii. superior alveolar—to upper teeth
3. *Mandibular—over the mandible and in front of the ear*
   i. buccal—to the skin over the mandible and cheek
   ii. auriculotemporal—to the auricle, eardrum, and temporal skin
   iii. inferior alveolar—to the lower teeth—mental branch via the mental foramen to the chin

B. Motor

1. *Facial nerve (CN VII)—extends thru the stylomastoid foramen and then emerges at the level of the tragus thru the parotid gland to supply the muscles of expression.*
   The mandibular branch runs forward below the angle of the mandible under the platysma, then goes to the corner of the mouth.
   Extracranial surgical or inflammatory damage to the Facial nerve may result in Bell's palsy (inability to wrinkle the forehead, whistle, or make a display of the teeth).
2. *The mandibular nerve (CN V)—to the muscles of mastication*

Facial Landmarks

I. BONY (1)

A. Mandibular

1. *Head of mandible—below zygoma and in front of the tragus*
2. *Coronoid process—can be palpated in front of the masseter and below the zygomatic arch*
3. *Articular fossa—can be felt with the mouth open*
4. *Ramus—between the masseter and medial pterygoid muscles*
5. *Angle—attachment for the internal (or medial) pterygoid which elevates the mandible*

B. Mental Foramen—lies in a line vertically below the supraorbital foramen and the infraorbital foramen

C. Infraorbital Foramen—1 inch below the inferior orbital margin (directly below the supraorbital foramen)

D. Glabella—connects superciliary arches

E. Supraorbital Notch (Foramen)—on the supraorbital margin about 1 inch from the median line

F. Zygomatic Arch—may be palpated from the front of the ear to below the eye

II. MUSCLES—which may be palpated with the teeth clenched

A. Temporal (48)—can be palpated in the temple

B. Masseter (55)—can be palpated near the angle of the jaw

III. ARTERIES (17)—pulsations which may be palpated

A. Facial Artery—pulsation felt as it crosses the lower margin of the mandible, at the anterior border of the masseter (1 inch in front of the angle of the mandible)

B. Superficial Temporal Artery—in front of the auricle

IV. FACIAL NERVE (19, 55)—emerges from the stylomastoid foramen, which is approximately 1 inch deep, to middle of the anterior border of the mastoid process. It courses anteriorly thru the parotid gland where it begins to divide into 5 branches.

V. PAROTID DUCT (55)—opens on the papilla opposite upper second molar tooth—can be felt as it crosses the masseter about 0.5 cm below the zygoma

Nasal Cavities

The nasal cavities are the posterior continuations of the nostrils. They extend to the choanae, the openings of the nasal cavities, into the nasopharynx.

I. WALLS (32–34)

A. Roof

1. *Anterior—frontal and nasal bones and cartilages*
2. *Central (Intermediate)—cribriform plate of ethmoid—lies below the cranial fossa. The central wall is perforated by olfactory nerves and ethmoidal vessels.*

B. Floor—palatine process of the maxilla and part of palatine bone

C. Medial Wall (Septum) (Columna)—formed by the vomer, the perpendicular plate of the ethmoid, and the septal cartilage

D. Lateral Wall—the 3 conchae

II. CONCHAE AND MEATI (32)—Conchae are elevations (turbinates) on the lateral wall of the nasal cavity. Meati are passageways for air and are located below the conchae.

A. Superior Meatus—The superior concha overlies the superior meatus and lies beneath the cribriform plate.

1. *Posterior ethmoidal cells open into the superior meatus.*

B. Middle Meatus—The middle concha overlies the middle meatus.

1. *The **hiatus semilunaris** is the opening for the frontal sinus, the maxillary sinus, and the anterior and middle ethmoid sinuses.*
2. *The lateral wall of the middle meatus contains a bulge (the ethmoid bulla).*

C. Inferior Meatus—lies below the inferior concha

The inferior concha projects medially into the nasal cavity from the sidewall and may contact the septum if it becomes swollen.

1. *The **nasolacrimal duct** opens into the inferior meatus.*
2. *The opening for the eustachian (auditory) tube is behind the inferior concha.*

III. MUCOUS MEMBRANE (32–40)

A. Divisions

1. *Olfactory (Superior) portion—composed of nonciliated columnar epithelium—located above the middle concha (i.e., in the upper central portion of the nasal cavity)*

2. *Respiratory (Inferior) portion—composed of pseudostratified, ciliated, columnar epithelium with goblet cells. The respiratory portion lines the remainder of the nasal cavity from the antrum posteriorly.*

B. Arteries (36)

1. *Sphenopalatine—from the maxillary from the external carotid—supplies the posterior septal and lateral walls*
2. *Ethmoidal—from the ophthalmic from the internal carotid—supplies the superior portion of the mucosa and nasal septum*
3. *Greater palatine—from the maxillary from the internal carotid—also supplies the nasal septum*

C. Veins (There is a submucous venous plexus.)

1. *Ethmoidal—to the superior sagittal sinus*
2. *Nasal—to the ophthalmic veins to the cavernous sinus*
3. *Sphenopalatine—to the pterygoid plexus to the cavernous sinus*

D. Lymphatics—to the cervical nodes

E. Nerves (37–40)

1. *Olfactory (CN I)—supplies the olfactory area before it passes thru the cribriform plate and pierces the meninges to enter the olfactory bulb*
2. *Trigeminal (CN V)—supplies the respiratory (lower) portion of the nasal mucosa*
   i. nasopalatine branch—to the septum
   ii. nasociliary branch of the Ophthalmic
3. *Parasympathetics—via CN VII (Facial) and the greater superficial petrosal nerve to the pterygopalatine ganglion (located in the pterygopalatine fossa and below the foramen rotundum) and thence to nasal mucous membranes*
4. *Sympathetics—from T1 and T2 spinal nerves to the superior cervical sympathetic ganglion and thence to nasal mucous membranes via the plexuses of the external carotid and maxillary arteries to the descending palatine and posterior nasal nerves*

Paranasal Sinuses

Paranasal sinuses (42, 43) are air-filled cavities named according to the bones in which they are found (frontal, maxillary, ethmoid, and sphenoid). They, plus the nasolacrimal duct, drain into the nasal cavities.

I. FRONTAL

A. Located behind the superciliary ridge of the frontal bone

B. Drain into middle meatus (via hiatus semilunaris)

C. Arteries—supraorbital from the ophthalmic from the internal carotid

D. Innervated by supraorbital nerves (of Ophthalmic of CN V)

II. MAXILLARY (Antrum)—largest of the paranasal sinuses

A. Located in the maxillary bone lateral to the nasal cavity

B. Drain into middle meatus

C. Arteries—supraorbital and superior alveolar

D. Innervated by alveolar nerves (from the Maxillary)

1. *The posterior superior alveolar nerve innervates the molar teeth.*

III. ETHMOID

A. Located between the orbit and nasal cavities

B. Anterior and Middle Ethmoidal sinuses drain into the middle meatus (via hiatus semilunaris).

C. Posterior Ethmoid—drains into superior meatus

D. Arteries—ethmoidal from the ophthalmic from the internal iliac

E. Innervated by the nasociliary nerves (from the Ophthalmic)

IV. SPHENOID

A. Located in the sphenoid bone (behind the orbit) above and behind the nasal cavities—separated by a thin bony plate extending from the pituitary gland and optic chiasma toward the nasal cavity

B. Drain into the sphenoethmoid recess (located above the superior concha in the nasal cavity)

C. Supplied by ethmoidal vessels and nerves

Oral Cavity

The oral cavity is bordered by the dental arches.
The portion between the cheeks and lips is called the vestibule.

I. BOUNDARIES (45, 46)

A. Anterior—lips—skin and mucous membrane

B. Lateral—the cheeks with buccinator muscle and fascia between

C. Superior

1. *Hard palate—formed of parts of the maxillary and palatine bones—underside covered by mucous membrane*
2. *Soft palate—covered by mucous membrane*
   i. continuous laterally with the palatoglossal arch (anterior pillar of the tonsil) and the palatopharyngeal arch (posterior pillar of the tonsil). The arches contain muscles.
   ii. The posterior end of the soft palate is called the uvula.
   iii. muscles
      a. palatoglossus and palatopharyngeus—contribute to closing off the mouth from the pharynx
      b. levator and tensor veli palatini—elevate the soft palate

D. Posterior—opening to the pharynx

E. Inferior—muscles in the floor of the mouth

II. OPENINGS OF THE DUCTS OF THE SALIVARY GLANDS (45)

A. Parotid—ducts open opposite upper second molar

B. Sublingual—ducts open around tongue

C. Submandibular—ducts open into area next to the lingual frenulum

III. TONGUE (52–57, 67)—(covered by papillated stratified squamous epithelium)

A. Muscles

1. *Intrinsic—muscles cross at angles providing tongue mobility (e.g., superior, inferior, longitudinal, and transverse)*
2. *Extrinsic—change the shape and position of the tongue*
   i. hyoglossus—from the hyoid bone to the side of the tongue—pulls the tongue down

ii. styloglossus—from the styloid process, between external and internal carotids, to the side of the tongue—pulls the tongue back

iii. palatoglossus—from the soft palate, then anterior to tonsil to the side of the tongue—closes off mouth from pharynx

iv. genioglossus—from the mandible to the tongue—projects the tongue forward and depresses the tongue

B. Artery—lingual from the external carotid

➢ C. Lymphatics (67)—to submental and submandibular nodes. (Lymph follicles at the root of the tongue are called the lingual tonsil.)

➢ IV. NERVES TO THE ORAL CAVITY (56, 129)

A. Sensory

1. *CN V (Mandibular) and CN IX (Glossopharyngeal)—sensory to anterior tongue (cell bodies in trigeminal ganglion)*

2. *CN VII (Facial)—taste to anterior tongue*

3. *CN IX (Glossopharyngeal)—taste to posterior tongue*

B. Motor

1. *CN XI (thru CN X)—innervates the palatal muscles*

2. *CN XII (Hypoglossal)—motor to the tongue*

➢ V. TEETH (50, 51)—32 permanent—(20 "baby teeth")

A. Groups

1. *Incisors—cutting teeth—1 root*

2. *Canines (Cuspids)—come to a point—1 root*

3. *Premolars (Bicuspids)—grinding surface on crown*

4. *Molars—4 cusps—larger grinding surface—2 or 3 roots*

B. Nerves—from the Trigeminal

1. *Upper teeth—Maxillary nerve*

2. *Lower teeth—Mandibular nerve*

Salivary Glands

I. PARAOTID (19, 55)—located between mastoid process and the ramus of the mandible. The parotid duct extends from the anterior border of the gland to cross the masseter muscle and then turn medially to go thru the buccinator muscle and open opposite the upper second molar. The external carotid artery bifurcates into the superficial temporal and maxillary in the parotid gland. The facial nerve goes thru the parotid to innervate the muscles of expression.

A. Artery—parotid from external carotid

B. Vein—retromandibular which drains into the external jugular

C. Lymphatics (66)—located on the surface—to parotid nodes to cervical nodes

D. Nerves (128)—parasympathetics—via CN IX (Glossopharyngeal nerve), inferior salivary nucleus, tympanic nerve, superficial petrosal nerve, and otic ganglion (located next to the Mandibular nerve as it exits the foramen ovale) and thence via the auriculotemporal nerve to the parotid gland

II. SUBMANDIBULAR (55)—located along the body of the mandible. Its duct begins between the mylohyoid and hyoglossus muscles and opens into orifices next to the lingual frenulum (45).

A. Artery—submandibular from facial from the external carotid

B. Vein—submental into the anterior facial into the internal jugular

C. Lymphatics—to submandibular nodes to deep cervical nodes

D. Nerves—parasympathetics via CN VII (Facial nerve), superior salivary nucleus, corda tympani, and submandibular ganglion (located between the submandibular gland and the hyoglossus muscle) and thence to the submandibular and sublingual glands. The lingual nerve has a branch to the submandibular gland.

III. SUBLINGUAL (55)—located in the floor of the mouth between the mandible and the genioglossus. The sublingual duct openings are located around the lingual frenulum.

A. Arteries

1. *Sublingual from the lingual from the external carotid*

2. *Submental from the maxillary from the external carotid*

B. Veins—lingual and submental veins drain into the anterior facial and thence into the internal jugular

C. Lymphatics—submental to submandibular nodes

D. Nerves—parasympathetics via CN VII (Facial nerve)

Mandible

I. MANDIBLE BONE (10)

A. Body—has a symphysis

1. *Lower portion—has a mental protuberance and tubercle*
2. *Middle portion—contains the mental foramen located midway between the borders of the mandible, 1 inch from the symphysis, and inferior to the second premolar tooth*
3. *Upper portion—alveolar*

B. Ramus

1. *Processes*
    - i. coronoid—anterior projection—attachment for temporal and mandibular muscles which elevate the mandible—separated from the condyloid by the mandibular notch
    - ii. condyloid—posterior projection
        - a. head (condyle) (capitulum)—articulates with the articular (mandibular) fossa of the temporal bone
        - b. neck
2. *Surfaces*
    - i. lateral (external)—rough and flat—insertion of the masseter muscle
    - ii. medial *(internal)*—uneven—contains mandibular foramen which transmits dental vessels and nerves

C. Angle of Mandible—located at the junction of the body and the ramus—attachment for the internal pterygoid which elevates the mandible

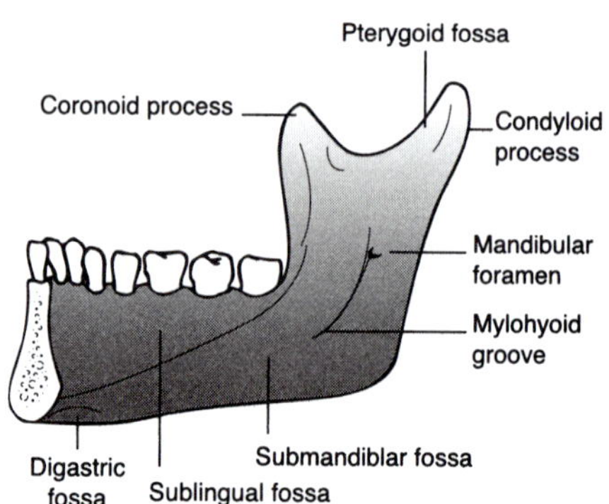

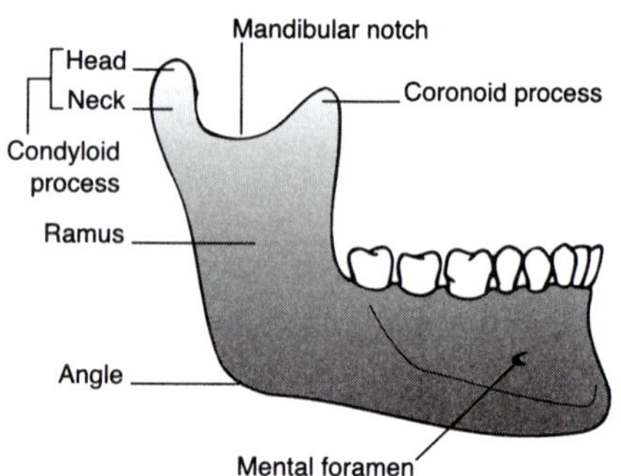

II. MUSCLES OF MASTICATION (48, 49)—supplied by the Mandibular nerve (except for the buccinator). The muscles insert on the mandible.

A. Elevate the Mandible and Close the Mouth—(Temporal, Masseter, and Internal Pterygoid)

1. *Temporal (Temporalis)*
    - i. from the temporal fossa to the coronoid process
    - ii. elevates and retracts the mandible
2. *Masseter*
    - i. from the zygomatic arch to the ramus and coronoid process
    - ii. elevates the mandible and clenches the teeth

3. *Internal (Medial) Pterygoid*

   i. from pterygoid lamina (plate) of the sphenoid to the angle of the mandible

   ii. elevates and protrudes the mandible

B. Depress the Mandible and Open the Mouth—(External Pterygoid and Anterior Belly of the Digastric)

1. *External (Lateral) Pterygoid*

   i. from the lateral plate of the sphenoid to the neck of the mandible and the disk of the mandibular joint

   ii. opens the mouth and protrudes the mandible

2. *Anterior Belly of the Digastric*

   i. from the mandible to the hyoid bone

C. Elevates the Tongue and Hyoid

1. *Mylohyoid*

   i. from the mandible to the hyoid bone

D. Aids in Mastication

1. *Buccinator—supplied by the Facial nerve*

   i. from the jaws near the lower teeth to lips

III. TM JOINT (11)—a synovial joint. The surfaces are separated by an articular disk. The temporal, masseter, and pterygoids act on the joint.

A. Articular Disk—(fibrocartilage)

1. *Attached to the capsular ligament*
2. *Upper surface fits into mandibular fossa and the eminence of the zygomatic process*
3. *Lower surface concave and fits into head of the mandible*
4. *The disk produces separate spaces within the joint.*

B. Ligaments—form a fibrous capsule

1. *TM—over the lateral aspect of the mandible*
2. *Sphenomandibular—medial to the joint*
3. *Stylomandibular—formed from deep fascia*

Neck

Page numbers in black. Netter plate numbers in green.

General

I. AREAS OF THE NECK

A. Triangles—separated by the sternocleidomastoid muscle

1. *Anterior—between the sternocleidomastoid, the mandible, and the median line*

2. *Posterior—between the sternocleidomastoid, the trapezius, and the middle of the clavicle*

B. Thoracocervical Area (Root)

➢ II. FASCIA (30)

A. Superficial—lies superficial to the platysma

B. Deep

1. *Superficial layer of deep fascia—lies between the superficial fascia and the muscle layer. Above the sternum it divides into two layers to form the suprasternal space.*

2. *Pretracheal—lies deep and encloses the infrahyoid muscles and the thyroid, trachea, and esophagus*

3. *Prevertebral—lies between the prevertebral muscles and the pharynx and esophagus. Prevertebral fascia extends from the skull to the superior mediastinum and forms the floor of the posterior triangle.*

4. *Carotid sheath—contains common and internal carotid arteries, internal jugular vein, and the Vagus nerve. The thoracic duct and inferior thyroid artery are posterior to the carotid sheath. The carotid sheath is a lateral extension of the pretracheal fascia.*

➢ III. COMPARTMENTS (30)—formed by deep cervical fascia

A. Visceral—located between the prevertebral and pretracheal fascia and the carotid sheath

1. *Retropharyngeal (retrovisceral) space—connects with the posterior mediastinum*

2. *Pretracheal space—anterior to the esophagus and around the trachea and thyroid*

B. Muscular—contains infrahyoid muscle

C. Suprasternal—contains the internal jugular veins

D. Vascular Fascial Compartment (Carotid Sheath)—contains the carotid arteries, internal jugular vein, Vagus nerve (located in the carotid sheath between the internal jugular vein and the internal and common carotid arteries), CN XII, and CN XI. The proximal portion of the Hypoglossal lies between the internal jugular vein and the internal and common carotid arteries.

➢ IV. BONES (12–14)—The spinous processes are joined by the ligamentum nuchae.

A. Cervical Vertebrae—have a foramen transversum for the vertebral artery, vein, and sympathetic nerves

1. *C1—Atlas—has no body or spine but 3 synovial joints. Upper facets articulate with the occipital condyles next to the foramen magnum.*
2. *C2—Axis—thickest and strongest—has a thick spine and bifid spinous process—has the odontoid process*
3. *C3 to C6—bifid spinous processes*
4. *C7—The anterior root is large and sometimes forms a cervical rib.*

B. Hyoid Bone—connects to the tongue and helps support the larynx

1. *Body—anterior (palpable)*
2. *Horns (Cornua)—laterally—greater (2) and lesser (2)*

➢ V. MUSCLES (22–25)

A. Sternocleidomastoids—separate anterior and posterior triangles. They extend from the manubrium and medial clavicles to the mastoid prominence of the temporal bone and are innervated by C2 and C3 and the spinal root of CN XI. One muscle can turn the head to the side.

Both sternocleidomastoids can raise the thorax slightly to increase its capacity during forced inspiration. (They may be injured during a breech vaginal delivery.)

B. Platysma—superficial to the sternocleidomastoid. It extends from the superficial upper thorax and the skin over the pectoralis major and deltoid muscles to the mandible and is innervated by the cervical branch of Facial nerve (CN VII). The platysma tenses (draws) the skin of the neck.

C. Scalene Muscles

1. *From cervical vertebrae to ribs 1 and 2. They flex the neck laterally.*
2. *Innervated by cervical nerves*

D. Digastrics—raise the hyoid bone and the base of the tongue

1. *Anterior belly from mandible to hyoid bone—innervated by a branch of the inferior alveolar nerve*

2. *Posterior belly from mastoid process to hyoid bone—innervated by CN VII*

VI. ARTERIES (28, 29)

A. Subclavian—in the root of the neck

B. Transverse Cervical and Inferior Thyroid—from the thyrocervical trunk

C. Deep Cervical—from the costocervical trunk from the subclavian

D. Common Carotid—Internal Carotid—External Carotid

E. Vertebral—supplies the deep muscles and spinal cord

VII. VEINS (26, 64)

A. Internal Jugular Vein—the largest vein in the neck.
It is a continuation of the sigmoid sinus. The internal jugular drains the brain and superficial parts of the face and neck and empties into the brachiocephalic vein. (Above the clavicle it is located just lateral to the carotid artery.)

1. *Common facial—enters the internal jugular near the hyoid bone*

   i. anterior facial

      a. angular—drains frontal and lateral nasal areas

      b. labial

      c. palpebral

      d. deep facial—drains the pterygoid venous plexus which drains the area served by the internal maxillary artery

   ii. posterior facial—formed by superficial temporal and internal maxillary veins

2. *Lingual—drains the tongue*

3. *Pharyngeal*

4. *Superior and middle thyroid—has a branch to the larynx*

B. External Jugular Vein—formed by the posterior auricular and the posterior facial—drains the scalp and face. In the neck it is located between the sternocleidomastoid and platysma muscles. It drains into the subclavian vein.

VIII. LYMPH NODES (66)

A. Superficial Cervical Nodes—drain into the deep nodes

B. Deep Nodes—The deep cervical nodes are located near the carotid artery and the internal jugular vein. They may be categorized as superior and inferior or medial and lateral.

IX. NERVES (27, 124)

A. Sympathetic Trunk—located between the carotid sheath and the prevertebral fascia—forms the 3 cervical ganglia posterior to the carotid sheath—contains sympathetic roots of spinal nerves and parasympathetic fibers

1. *Superior cervical ganglion (C1–C4)—postganglionics follow the carotid arteries—has pharyngeal, thyroid, external carotid plexus, and superior cardiac branches*

   HORNER'S SYNDROME is the result of damage to the ganglion and sympathetic pathway.

   Signs may include:

   A weakened superior tarsal (superior eyelid) muscle resulting in partial PTOSIS (partial eyelid droop)

   Unapposed parasympathetic activity to the pupillary sphincter resulting in MIOSIS (small pupil)

   Lack of facial sudomotor activity resulting in ANHIDROSIS (lack of sweat)

2. *Middle cervical ganglion (C5–C6)—has a middle cardiac branch*

3. *Inferior cervical ganglion (C7-T1)—has an inferior cardiac branch. When the inferior ganglion is joined to the first thoracic ganglion, the combination is called the* ***stellate ganglion.***

B. Motor Nerves

1. *A branch of the Facial nerve innervates the platysma muscle.*

2. *A branch of the Accessory nerve (CN XI) innervates the sternocleidomastoid muscle and the trapezius muscles.*

3. *Cervical plexus (C1–C4) has branches to several neck muscles—formed by the anterior rami of the first 4 cervical nerves.*

   i. C1 to geniohyoid and thyrohyoid

   ii. C1–C3 to sternohyoid, omohyoid, and sternothyroid

4. *Phrenic nerve—located anterior to the scalenus anterior muscle*

TABLE 3-1. MIDLINE OF THE NECK

STRUCTURE	LOCATION	IMPORTANCE
Hyoid bone	At the root of the tongue At the level of C3–C4 vertebrae	Greater cornu is close to many major structures Connected by muscles to the tongue, mandible, sternum, and scapula Upper border is near the lingual arteries Helps support the larynx—attachment for many laryngeal muscles
Thyrohyoid membrane	Between the hyoid bone and thyroid cartilage	Ligament which suspends the larynx from the hyoid bone
Thyroid cartilage	Between thyroid cartilage and cricoid cartilage At the level of C4 vertebra At the level of the bifurcation of the common carotid arteries	Laminae are united in the midline to form the laryngeal prominence "Adam's apple" Upper halves form the thyroid notch The posterior borders form cornua The cricoid can be felt below the thyroid cartilage Protects the larynx and vocal cords
Cricothyroid membrane	Between the thyroid and cricoid cartilages	May be a site for tracheostomy Below the thyroid cartilage and the thyroid gland
Cricoid cartilage	At the level of the lower end of the larynx At the level of C6 At the level where the common carotid artery is crossed by the omohyoid At the level of the junction of the pharynx and esophagus	Forms a ring encircling the larynx Supports the arytenoid cartilage (an attachment for the vocal cords)

Anterior Triangle

The anterior triangle is located between the sternocleidomastoid, the inferior border of the mandible, and the anterior median line of the neck. It contains the glandular structures of the neck.

I. SMALLER TRIANGLES IN THE ANTERIOR TRIANGLE (22, 23)

A. Muscular—contains the infrahyoid muscles and thyroid gland

B. Submental Triangle—located between the anterior belly of the digastric, the hyoid bone, and the midline—contains lymph nodes which drain the floor of the mouth, chin, and central lower lip

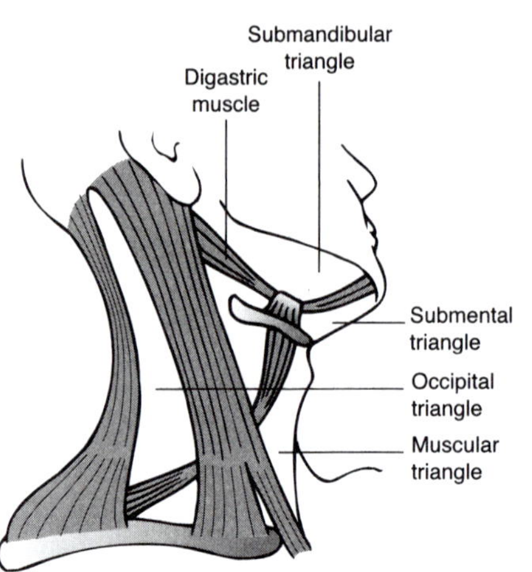

C. Submandibular (Digastric) Triangle—between the mandible and the two bellies of the digastric

1. *Anterior portion—contains the submandibular salivary gland, the lymph nodes, and the submandibular ganglion*
2. *Posterior portion—contains CN XI and the facial artery*

D. Carotid Triangle—located between the omohyoid, digastric, and sternocleidomastoid muscles

1. *Contains the bifurcation of the common carotid. In the carotid triangle, the external carotid lies medial and anterior to the internal carotid.*
2. *Contains the carotid sinus which is supplied by cranial nerves IX and X and sympathetics*
3. *The internal jugular vein—overlaps the carotid vessels*

II. MEDIAN LINE OF THE NECK (9, 71)

A. Suprahyoid Portion—the submental triangle

B. Hyoid Bone—contains 4 horns (2 pairs)

C. Infrahyoid Portion—extends from the inferior hyoid to the suprasternal (jugular) notch

III. MUSCLES (22–24, 47)

A. Muscle of Expression

1. *Platysma—from the mandible to the pectoralis major and deltoid fascia—innervated by the Facial nerve. The platysma tenses the skin of the neck.*

B. Muscles Which Move the Hyoid Bone

1. *Suprahyoid—mylohyoid, stylohyoid, geniohyoid, and digastrics—raise the hyoid bone*
2. *Infrahyoid (Strap muscles)—sternohyoid, thyrohyoid, omohyoid, and sternothyroid—lower the hyoid bone—innervated by C1, C2,*

*and C3 [via the **ansa cervicalis** which is a loop formed by the anterior primary rami of the upper (C1) and lower (C2–C3) roots]*

IV. ARTERIES (29)

A. Common Carotid—contains the carotid sinus

1. *Internal carotid—**no** branches in the neck—lies anterior to the sympathetic trunk and Vagus nerve and supplies the brain*
2. *External carotid*

   i. superior thyroid—to the parathyroid glands and the superior portion of the thyroid gland

   ii. facial (external maxillary)—arises in the carotid triangle

   iii. lingual (the lingual and facial may have a common trunk)

   iv. occipital—to posterior neck muscles and the dura

   v. pharyngeal

B. Branches of the Brachiocephalic

C. Branches of the Subclavian

1. *Vertebral arteries—enter the skull via the foramen magnum to supply the brain stem via the basilar artery*
2. *Inferior thyroid from the thyrocervical trunk*

V. VEINS (26)

A. Anterior jugular (a tributary of the external jugular)—drains the submental area

B. Internal Jugular—in the carotid sheath—joins the subclavian to form the brachiocephalic. Above the clavicle it is located just lateral to the common carotid artery **(a surgical danger area)**.

VI. CERVICAL LYMPH NODES (66)

A. Deep Cervical Nodes (15–30)—located along the carotid artery and internal jugular vein and divided into superior and inferior (supraclavicular) groups. The deep nodes receive lymph from the superficial nodes.

1. *Superior group—related to the internal jugular vein*

   i. submandibular—related to the internal jugular vein—receives lymphatics from the nose, lips, anterior tongue, and submandibular gland

ii. submental—related to the internal jugular vein and the mylohyoid muscle—drains the chin, the floor of the oral cavity, and the central lower lip

2. *Inferior (Supraclavicular) group—located near the subclavian vein—receive some lymphatics from the upper limb (apical axillary nodes) and the thoracic wall in addition to those from the head and neck*

B. Superficial Cervical Nodes—located along the external jugular vein—drain into the deep cervical nodes

VII. NERVES (124)

A. Parasympathetic—located in the carotid sheath

1. *Hypoglossal (CN XII)—in the carotid sheath*

2. *Vagus (CN X)—in the carotid sheath*

i. recurrent laryngeals

3. *Accessory (CN XI)—in carotid sheath and posterior triangle*

B. Sensory—C2–C4

VIII. BRANCHIAL (PHARYNGEAL) ARCHES—mesenchymal derivatives which become separated into clefts

A. First Arch (Mandibular)—derivatives include the muscles of mastication—innervated by the mandibular branch of CN V

Abnormal formation is associated with facial deformities such as cleft palate.

B. Second Arch (Hyoid)—derivatives include the muscles of expression—innervated by CN VII

BRANCHIAL CLEFT—Remainders of the 2nd branchial arch may result in a cyst (more common in adults) or a fistula (more common in children). The fistula lies along the carotid sheath and is close to the internal jugular vein. The opening is located between the middle and lower thirds of the sternocleidomastoid muscle.

Failure to remove the epithelial lining of the tract may result in recurrence.

C. Third Arch—contributes to the formation of the stylopharyngeus muscle—innervated by CN IX

D. Fourth to Sixth Arches—contribute to the formation of the pharyngeal constrictors and the intrinsic laryngeal muscles—innervated by branches of CN X

Posterior Triangle

The posterior triangle is located behind the sternocleidomastoid, anterior to the trapezius, and above the clavicle. The trapezius and sternocleidomastoid meet at the apex of the triangle.

➢ **I. SMALLER TRIANGLES IN THE POSTERIOR TRIANGLE (164)—separated by the omohyoid muscle**

A. Occipital—contains the accessory nerve

B. Subclavian (Supraclavicular)—contains the subclavian artery

➢ **II. MUSCLES (163, 164)**

A. Levator Scapulae—lifts the scapula

B. Scalenus Medius and Scalenus Posterior—flex the neck

C. Splenius Capitis—flexes neck and head

D. Platysma—from the mandible to the fascia over the deltoid and pectoralis major

➢ **III. ARTERIES (164)**

A. Transverse Cervical—supplies muscles in the posterior triangle and scapular muscles—from the thyrocervical trunk

1. *Subscapular—supplies scapular muscles*

B. Occipital—passes thru the apex of the triangle

➢ **IV. VEINS (64)—The external jugular vein drains the scalp, face, and superficial portions of the neck.**

V. LYMPH NODES—superficial cervical

➢ **VI. NERVES IN THE POSTERIOR TRIANGLE (121, 164)**

A. Sensory

1. *Lesser occipital—to the skin*
2. *Greater auricular—to the skin over the parotid, mastoid, external ear, and ear canal*
3. *Supraclavicular—to skin of upper posterior shoulder area*
4. *Transverse cervical—to the front and side of the neck*

B. Motor

1. *Cervical plexus (C1–C4)—C1 to the infrahyoid muscles—C1–C3 to the sternohyoid, omohyoid, and sternothyroid—Phrenic (C3–C5) to the diaphragm*
2. *Accessory nerve (CN XI)—supplies the sternocleidomastoid and trapezius*

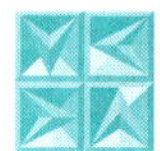

Thoracocervical Region

The thoracocervical region is sometimes called the *root* of the neck. Blood vessels, nerves, and lymphatics pass thru this region.

I. BOUNDARIES

A. Anterior—manubrium

B. Posterior—T1 vertebra

C. Lateral—first ribs

II. ARTERIES (28)

A. Brachiocephalic Trunk—2 inches long—arises posterior to the manubrium

B. Subclavians—arise posterior to the sternoclavicular joint—the right from the brachiocephalic trunk, the left from the arch of the aorta. The subclavians pass thru the root of the neck between the clavicle and 1st rib to accompany the brachial plexus into the apex of the axilla.

1. *Vertebral arteries—supply the spinal cord and part of the brain—enter the skull thru the foramen magnum—form the basilar artery in the brain*

2. *Internal thoracic arteries—to intercostal spaces*

3. *Thyrocervical trunk—begins medial to scalenous anterior*

   i. inferior thyroid—to the inferior thyroid and parathyroid

   ii. suprascapular—to the muscles around the scapula

   iii. transverse cervical—to the posterior triangle muscles

III. VEINS (28) —(The brachiocephalic vein is formed by the subclavian and the internal jugular.)

A. Subclavian—passes between the clavicle and 1st rib—lies anterior to the subclavian artery and is a continuation of the axillary vein. The vertebral veins drain into the right subclavian.

1. *External jugular*

   i. anterior jugular—begins in the anterior triangle (submental area) to travel beneath the sternocleidomastoid muscle and drain into the external jugular vein

B. Internal Jugular—located in the carotid sheath. It joins with the subclavian to form the brachiocephalic vein.

IV. LYMPHATICS (66)

A. Thoracic Duct—receives all lymphatics below the diaphragm and the left half of the thorax—drains **into** the left brachiocephalic vein or left subclavian or left internal carotid

1. *At the level of the cervicothoracic junction, the thoracic duct lies on the prevertebral fascia.*
2. *At the level of the T1 vertebra, the thoracic duct loops over the subclavian artery.*
3. *It lies posterior to the carotid sheath, in front of the vertebral vessels, and lateral to the esophagus.*
4. *On the left side, the left jugular, left subclavian, and left mediastinal lymphatics drain into the thoracic duct.*
5. *Virchow's node—located near the head of the sternocleidomastoid muscle. Its efferents drain into the thoracic duct.*

B. Right Lymphatic Duct—The right lymphatic duct drains the right upper limb, right thorax, and the right side of the head and neck. It drains **into** the right internal jugular or right subclavian vein.

1. *The right lymphatic duct drains the right upper limb, the right thorax, and the right side of the head and neck.*
2. *The right subclavian and right jugular trunks drain into their respective veins* ***or*** *the right lymphatic duct.*
3. *The right mediastinal trunk may drain directly into the brachiocephalic vein.*

V. NERVES (124)

A. Parasympathetic

1. *Recurrent laryngeal nerve—supplies laryngeal muscles except the cricothyroid*
   i. On the left side, the recurrent laryngeal enters the thorax in the carotid sheath and then loops under the aortic arch and behind the ligamentum arteriosum, to ascend in the neck alongside the trachea and esophagus.
   ii. On the right side, the recurrent laryngeal may leave the carotid sheath at a higher level.
2. *Phrenic (C3, C4, C5)—supplies the diaphragm*

B. Sympathetic Trunks and Ganglia

1. *Postganglionic nerves proceed on as cervical spinal nerves.*
2. *Visceral branches proceed directly without synapse in ganglia.*

➢ VI. **THORACIC OUTLET SYNDROME (TOS) (173) Compression of the subclavian vessels, or, more commonly, the lower brachial plexus, may be caused by the clavicle or 1st rib or by the scalenus muscle.**

1. ***Neurogenic TOS—Compression of a portion of the brachial plexus (inferior trunk) may produce shoulder and arm pain and ulnar nerve changes including thenar muscle weakness.***
2. ***Vascular TOS—Compression of the subclavian vein or artery by the 1st rib or a cervical rib may result in thrombosis and emboli.***

Thyroid

The thyroid is a two-lobed endocrine gland connected by an isthmus and surrounded by pretracheal fascia. This fascia extends to the trachea and larynx. The gland is located 1 inch below the laryngeal prominence in the arch of the cricoid.

The thyroid develops from the thyroglossal duct. A patent duct may form a thyroglossal duct cyst usually found in the midline above the thyroid cartilage. The duct is a retained connection between the former foramen cecum and the area anterior to the hyoid bone.

I. DIVISIONS (68)

A. Lateral Lobes (2)—Relationships:

1. *Anteriorly—are the infrahyoid and sternocleidomastoid muscles*
2. *Posteriorly—are the esophagus, lower pharynx, and carotid sheath. The parathyroids lie on the posterior surface of the gland.*
3. *Posteromedially—is the recurrent laryngeal nerve*

B. Isthmus—located in front of the 2nd, 3rd, and 4th tracheal rings

C. An Accessory Lobe (Pyramidal) sometimes extends upward from the isthmus. It is another remnant of the thyroglossal duct.

II. ARTERIES (63, 69, 70)

A. Superior Thyroid—from the external carotid

B. Inferior Thyroid—from the subclavian and thyrocervical trunk. It lies close to the recurrent laryngeal nerve.

C. Thyroid Ima—an artery, sometimes present, from either the brachiocephalic or aortic arch. When present, it may extend over the trachea to the inferior portion of the thyroid.

III. VEINS (68)

A. Superior Thyroid Vein—drains into the internal jugular vein

B. Middle Thyroid Vein—drains into the internal jugular vein

C. Inferior Thyroid Vein—runs anterior to the trachea and drains into the brachiocephalic vein

IV. LYMPH NODES (66)—Pretracheal and prelaryngeal nodes to the deep cervical nodes

V. NERVES (124)—via cervical sympathetic ganglia. Postganglionic fibers are vasomotor.

Parathyroid

The parathyroids are 4 pea-sized glands on the posterior aspect of the thyroid, within its capsule, and in the pretracheal fascia. (There may be as few as 2 or as many as 8 glands, and they may even be located outside of the thyroid capsule.)

I. RELATIONSHIPS (70)

A. Superior Parathyroids—frequently located close to the pharyngoesophageal junction about the middle of the thyroid. They usually lie superior to the inferior thyroid artery and posterior to the recurrent laryngeal nerve.

B. Inferior Parathyroids—frequently located near the lower end of the thyroid. The inferior parathyroids lie close to the recurrent laryngeal nerve.

II. ARTERIES (70)

A. Superior Thyroid arteries (from the external carotid) supply the superior parathyroids.

B. Inferior Thyroid arteries (from the thyrocervical trunk) supply the inferior parathyroids.

III. VEINS—drainage to the thyroid veins

IV. LYMPHATICS—along with those of the thyroid gland to the pretracheal, to the prelaryngeal, or directly to deep cervical nodes

V. NERVES (70)

A. Parasympathetics via the recurrent laryngeal nerve

B. Sympathetic trunks and ganglia

Thymus

The thymus is derived from entodermal diverticula. It is located between the sternum and great vessels in the region of the anterior and the superior mediastina of the thoracocervical region. The purpose of the thymus is to process lymphocytes into immunocompetent T cells.

I. RELATIONSHIPS (200, 218, 219)

A. Anteriorly—are the sternum and sternohyoid muscles

B. Posteriorly—are the pericardium, ascending aorta, and trachea

II. ARTERIES—from the internal thoracic and inferior thyroid

III. VEINS—internal thoracic and thyroid

IV. LYMPHATICS—to parasternal and tracheobronchial nodes

V. NERVES—Vagus (CN X)

Larynx

The larynx (voice organ) protects the opening of the air passage. It consists of cartilages connected by joints and elastic fibers. The larynx is located between the trachea and the tongue and extends from the epiglottis to the cricoid.

I. AREAS (75)

A. Vestibule—the area (cavity) superior to the vestibular folds (false cords). It extends from the laryngeal inlet to the vestibular folds.

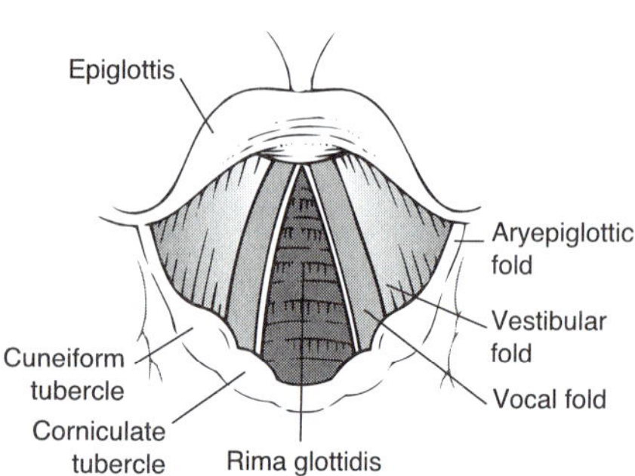

B. Glottis—the true and false cords plus the recess (rima) between them

1. *Rima (Rima glottidis)—a fissure (opening) between the vocal folds and the vocal processes of the arytenoid cartilage. It is the narrowest part of the larynx.*
2. *Folds*
   i. vestibular folds (false vocal cords)—the mucous membrane enclosing the vestibular ligament
   ii. vocal folds (true vocal cords)—consist of vocal ligaments, accompanying muscle fibers, and their covering mucous membrane. The vocal folds are covered by stratified squamous epithelium.

C. Infraglottic Area—the area from the vocal folds to the first tracheal ring

II. CARTILAGES (71, 75)—3 single and 3 paired

A. Thyroid—consists of two plates (laminae) which are fused at the median line of the neck. The prominent portion is called the "Adam's apple."

1. *Cornui*
   i. superior cornu—for the thyroid membrane (lateral hyothyroid ligament)
   ii. inferior cornu—articulates with the cricoid cartilage

B. Cricoid (single)—2 plates—shaped like a signet ring—forms the lower walls of the larynx—attached to the thyroid cartilage by the cricothyroid ligament

C. Arytenoid (paired)—located at the upper posterior border of the cricoid cartilage and attached to the aryepiglottic fold

1. *Has a lateral projection ("muscular process")*
2. *Has an anterior projection ("vocal process") which supports the vocal ligament*

D. Corniculate (paired)—located in the aryepiglottic fold

E. Cuneiform (paired)—in the aryepiglottic fold—located to the sides of the arytenoids

F. Epiglottis (single)—over the superior aperture of the larynx. The epiglottis forms the anterosuperior wall of the larynx and closes off the larynx during swallowing.

Paralysis of the recurrent laryngeal nerve can impair swallowing.

1. *Connected to the thyroid cartilage by the thyroepiglottic ligament*
2. *Connected to the hyoid cartilage by the hyoepiglottic ligament*
3. *The piriform recess is under the epiglottis.*

III. LIGAMENTS (71)

A. Extrinsic

1. *Thyrohyoid (membrane)—connects the thyroid cartilage with the hyoid bone*
2. *Cricotracheal—connects the cricoid cartilage with the 1st tracheal ring*
3. *Hypoepiglottic—from the hyoid bone to the epiglottis*
4. *Cricothyroid—from the cricoid to the thyroid cartilage*

B. Intrinsic

1. *Vestibular ligament*
   i. the inferior margin of the quadrangular membrane (connective tissue between the arytenoid and epiglottis)
      a. The aryepiglottic fold extends along the superior border of the quadrangular membrane.
   ii. The vestibular ligament is covered by the vestibular fold of mucous membrane (false vocal cord).
2. *Vocal ligament*
   i. extends from the thyroid cartilage to the vocal process of the arytenoid
   ii. covered by the vocal fold
3. *Articular capsule—around the articulation of the inferior part of thyroid and cricoid cartilage*
4. *Thyroepiglottic ligament—extends from the epiglottis to the thyroid cartilage (near the superior thyroid notch)*

IV. MUSCLES (72, 73)

A. Extrinsic—located between the larynx and surrounding parts. They move the larynx as a whole and support the larynx. Included are muscles attached to the hyoid bone. They are sometimes classified as elevators or depressors.

B. Intrinsic—voluntary muscle fibers. They are limited to the larynx proper and are attached to the cricoid, arytenoid, and thyroid cartilages.

The intrinsic laryngeal muscles control phonation.

1. *Thyroarytenoid—relax vocal folds*
2. *Thyroepiglottic—widen the inlet of the larynx*
3. *Cricoarytenoid—open and close the folds*

V. ARTERIES (70)

A. Superior Laryngeal Arteries—from the superior thyroid from the external carotid

B. Inferior Laryngeal Arteries—from the inferior thyroid from the thyrocervical trunk

VI. VEINS

A. To Superior Thyroid veins to the internal jugular veins

B. To Inferior Thyroid veins to the brachiocephalic veins

VII. LYMPHATICS—to nodes near the bifurcation of the common carotid arteries and to the deep cervical nodes

VIII. NERVES (68–70, 74, 223)

A. Parasympathetic—Vagus (CN X)

1. *Superior laryngeal*
   i. internal branch—sensory to the laryngeal mucosa superior to the vocal cords
   ii. external branch—to the cricothyroid muscle
2. *Recurrent laryngeal—supplies all laryngeal muscles except the cricothyroid—terminates as the inferior laryngeal nerve*
   i. inferior laryngeal—supplies the intrinsic muscles of the larynx

B. Accessory Nerve (CN XII)—innervates the intrinsic laryngeal muscles thru the laryngeal branches of the Vagus

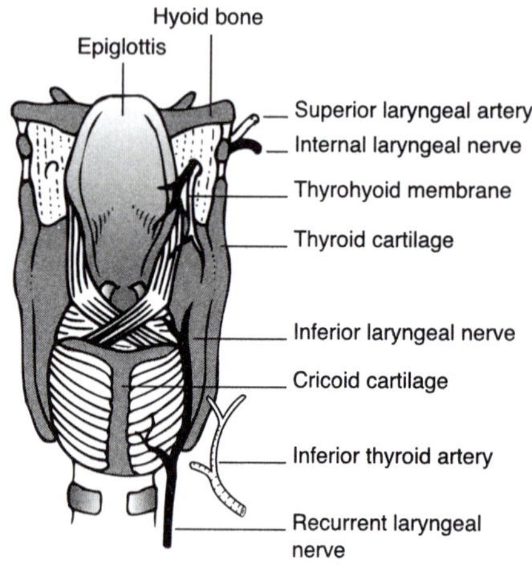

Pharynx

The pharynx is located posterior to the mouth and larynx. It extends from the base of the skull to the level of the cricoid cartilage.
The nasal apertures, pharyngotympanic tubes, larynx, and esophagus open into the pharynx.

I. DIVISIONS (57, 60)

A. Nasopharynx—respiratory

1. *Internal nares (choanae)*

i. auditory (eustachian) tube orifice—on the lateral wall

ii. pharyngeal tonsil—lymphoid tissue on the posterior wall

iii. pharyngeal recess—on the posterolateral wall

B. Oropharynx—digestive and respiratory pathway—located between soft palate and the base of the tongue

1. *Contains the tonsils (Palatine)*

C. Laryngopharynx—digestive pathway—located posterior to and connected to the larynx

1. *Piriform recess—on each side of the inlet of the larynx*

2. *Aryepiglottic fold—separates the piriform recess from the inlet*

3. *The thyroid and carotid sheath are located laterally.*

II. FASCIA (30)

A. Prevertebral—between vertebrae and pharynx

B. Pharyngobasilar—attached to the skull

C. Buccopharyngeal—on the buccinator muscle

➢ III. ARCHES (58)—contain muscles which decrease the opening between the nasopharynx and the oropharynx

A. Palatine Arch—mucosal folds at the sides of the posterior border of the soft palate

B. Palatopharyngeal (Posterior pillars)—extend from the pharyngeal wall to the soft palate near the uvula

C. Glossopalatine (Anterior pillars)—extend from the tongue to the soft palate

➢ IV. MUSCLES (59, 61, 62, 69)

A. Constrictors—The superior (related to the hyoid bone), middle, and inferior constrictors overlap. The pharyngeal constrictors surround the pharynx, change resonance, and propel food.

B. Elevators—stylopharyngeus (contains the Glossopharyngeal nerve), palatopharyngeus, and salpingopharyngeus

V. ARTERIES (63) —branches of the external carotid

A. Ascending Pharyngeal

B. Ascending Palatine of the facial

C. Descending (Greater) Palatine from the maxillary

VI. VEINS—pharyngeal venous plexus to the internal jugular

VII. LYMPH NODES (67) —superior group of the deep cervical

VIII. NERVES (65)

A. Sympathetic—from the superior cervical ganglion

B. Parasympathetic—CN IX, CN X, and CN XI are variously involved in the innervation of the pharyngeal musculature.

1. *Glossopharyngeal (CN IX)—innervates the stylopharyngeus*

2. *Vagus (CN X)—innervates the constrictors. The recurrent laryngeal innervates muscles in the aryepiglottic fold.*

C. Glossopharyngeal (CN IX)—carries sensory fibers from the walls of the pharynx

IX. TONSILS (57, 58) —localized lymphatic tissue

A. Palatine Tonsils—located between the palatine arches (in the tonsillar fossa) and extend to the soft palate—about 1 inch long—contain crypts lined with stratified squamous epithelium

1. *Arteries—the tonsillar branches of the facial, palatine, and lingual arteries*

2. *Veins—palatine drains into the common facial vein and the pharyngeal plexus*

3. *Nerves—Glossopharyngeal (CN IX)*

4. *Lymphatics—to the tonsillar node (jugulodigastric) with efferents to deep cervical nodes*

B. Nasopharyngeal Tonsils (Adenoids)—lymphoid tissue in the upper posterior wall of the pharynx near the opening of the eustachian tube

NOTE: The palatine, pharyngeal, and lingual tonsils form a tonsillar (Waldeyer's) ring.

Thorax

Chapter

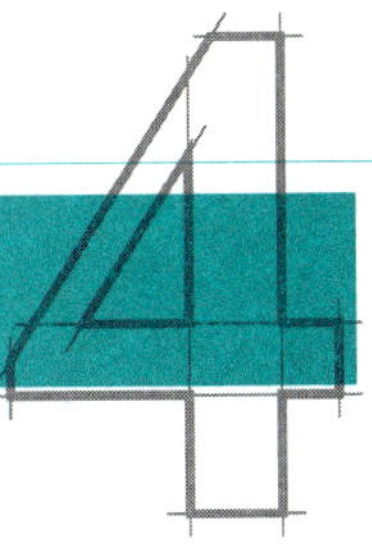

Page numbers in black. Netter plate numbers in green.

Thoracic Arteries

I. ASCENDING AORTA

A. Coronary Arteries

II. AORTIC ARCH (225)—begins behind the right edge of the sternum. It extends from the second costochondral junction to the level of T4 posteriorly.

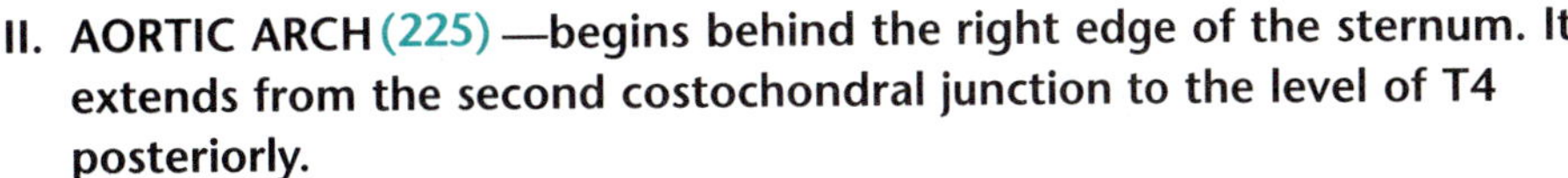

The dilation in an arterial aneurysm consists of all three arterial layers (intima, media, and adventitia). In a traumatic pseudoaneurysm, all three layers are broken.

A. Brachiocephalic (formerly called the Innominate)

1. *Right subclavian*
    i. internal thoracic (formerly the internal mammary)
        a. anterior intercostals—supply the upper anterior intercostal spaces, muscles, parietal pleura, and breast
        b. musculophrenic—supplies intercostal spaces 7 to 9 and pectoral muscles
        c. superior epigastric—supplies the diaphragm
2. *Right common carotid*
3. *Costocervical—supplies the two uppermost intercostal spaces*

B. Left Common Carotid

C. Left Subclavian—the 3rd branch of the aortic arch

1. *Internal thoracic (mammary)*
2. *Costocervicals*
3. *Vertebral—supplies part of the brain and spinal cord*
4. *Thyrocervical trunk*

III. DESCENDING AORTA (179, 225)—begins in left posterior mediastinum at the level of the sternal angle and T4

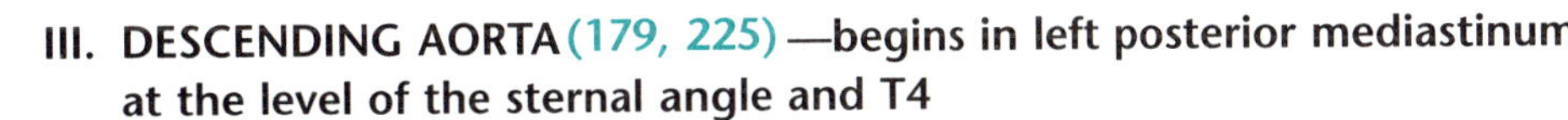

A. Posterior Intercostals—to the lower 9 intercostal spaces

B. Subcostals—below the 12th rib

C. Pericardials

D. Esophageals (4–5)

E. Bronchials—right and left, to the lungs

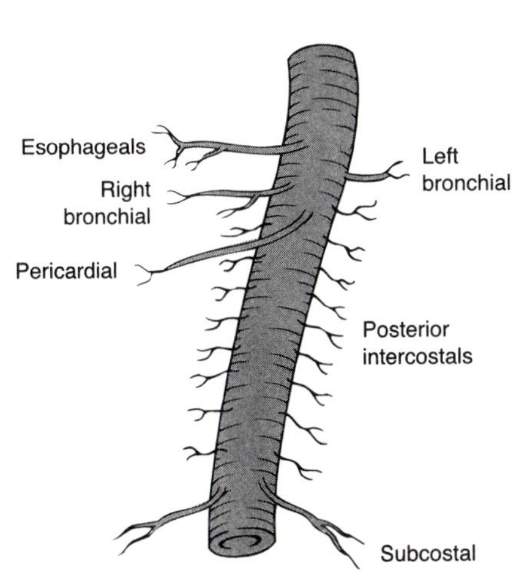

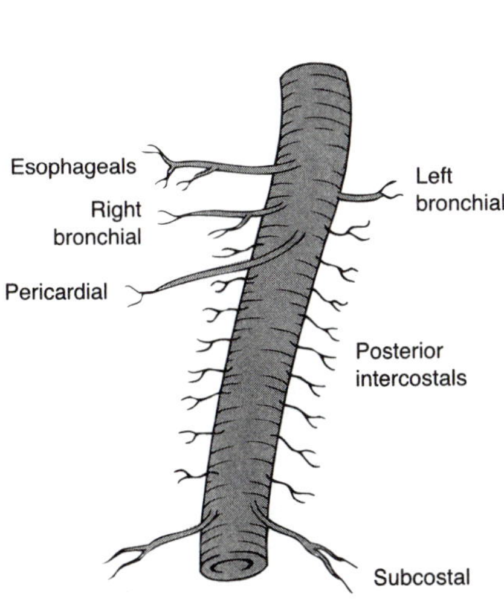

Nervous Systems

I. FUNCTIONAL CLASSIFICATION

A. Somatic Nervous System

1. *Afferents—convey sensory information*
2. *Efferents—involved in control of voluntary muscles*

B. Autonomic Nervous System (ANS)

1. *Sympathetic Nervous System (SNS)—stimulated under "stress"*
2. *Parasympathetic Nervous System (PNS)—"conserves body resources"*

II. FIBERS OF THE SOMATIC NERVOUS SYSTEM (166, 176, 179)—In the thoracic region, the ventral primary rami are termed "intercostal nerves." They extend into the intercostal spaces and innervate muscles, skin, and, parietal pleura via lateral and anterior branches. From T6 to T12, the nerves continue onto the anterior abdominal wall as far down as the umbilicus.

A. Afferents—transmit pain, touch, and temperature from the pectoral wall

B. Efferents—supply the muscles via ventral roots

III. FIBERS OF THE ANS (179, 198, 228)

A. Sympathetic (SNS)

1. *Preganglionic fibers*

   i. origination—in cell bodies in the spinal cord

   ii. course—they proceed out ventral roots thru anterior primary rami, then thru white rami communicans, then thru the sympathetic chain, then to the paravertebral ganglia. After entering the sympathetic chain, fibers may pass up or down. **Some** preganglionic fibers pass thru the sympathetic chain to form splanchnic nerves. The splanchnic nerves go down to supply the abdominal viscera. On their course, they pass under the medial arcuate ligament.

   iii. sympathetic trunk—has 12 ganglia each with a white and gray ramus which passes to a spinal nerve.

   iv. termination of preganglionic fibers

      a. paravertebral ganglia—sympathetic ganglia which form the left and right sympathetic chains (trunks) on the sides of the vertebral column **or**

      b. prevertebral ganglia—sympathetic ganglia for the splanchnic nerves

2. *Postganglionic fibers pass either to:*
   i. gray rami communicans to pectoral wall blood vessels, sweat glands, and muscles **or**
   ii. thoracic viscera after a synapse in one of the first 5 ganglia of the sympathetic trunk
      a. heart—SNS increases beat and dilates coronary arteries
      b. lungs—SNS dilates bronchi
      c. esophagus—SNS constricts blood vessels

B. Parasympathetic (PNS) Fibers

1. *Preganglionic—PNS fibers in the thorax are contained in the Vagus and terminate in cell bodies close to the organ.*
2. *Postganglionic—enter an organ*
   i. heart—PNS slows beat and constricts coronary arteries
   ii. lungs—PNS constricts bronchi

IV. VISCERAL AFFERENTS (215)

A. Esophagus—to T1–T5 via thoracic splanchnics

B. Heart—to T1–T4 via cervical and thoracic splanchnics

C. Diaphragm—to C3–C5 via the phrenic nerve

D. Pleura

1. *Visceral—via vagal and sympathetic fibers*
2. *Parietal—via intercostal nerves*

V. ENTERIC NERVOUS SYSTEM (ENS)

A. Relaxation of the Lower Esophageal Sphincter (LES) is caused by inhibitory neurons using Vasoactive Intestinal Peptide (VIP), Nitric Oxide (NO), and the influence of Cholecystokinin (CCK) on VIP and NO neurons.

B. Contraction of the LES is caused by the direct action of CCK.

Pectoral Region

I. DEFINITIONS

A. The Thorax (chest) is the area between the neck and abdomen.

B. The Osteocartilaginous Thoracic "Cage" encloses the thoracic cavity. It has superior and inferior apertures.

C. The Thoracic Cavity houses the thoracic viscera. It includes the pleural cavities and the mediastinum.

D. The Thoracic Wall consists of bones, muscles, and the parietal pleura around the thoracic cavity.

E. The Pectoral (Shoulder) Girdle consists of the scapula and clavicle.

F. Pleural Cavity—a "potential" space between the parietal and visceral pleurae

➢ II. BONY THORAX (170, 171)

A. Thoracic Vertebrae

1. *Spinous processes (spines)*
   - i. horizontally directed—T1, T2, T11, T12
   - ii. inferiorly directed—T3–T10
2. *Costal rib facets—on bodies and transverse processes of T1–T10*

B. Sternum

1. *Manubrium—joins the body at the manubrial-sternal joint at the sternal angle (angle of Louis).*
   NOTE: *The costal cartilage of the 2nd rib articulates with the sternum at the sternal angle. The sternal angle also marks the approximate level of the arch of the aorta, the termination of the azygos, the corina of the trachea, and vertebral level T4–T5. (514)*
   - i. notches
     - a. suprasternal (jugular)—(T2 vertebral level)
     - b. clavicular
     - c. notches for the 1st costal cartilages
2. *Body—articulations*
   - i. superior—manubrium
   - ii. lateral—2nd costal cartilage at the sternal angle, costal cartilage of ribs 3–6, and the 7th rib at the body-xiphoid articulation
   - iii. inferior—xiphoid (vertebral level T9)
3. *Xiphoid (lies at the level of T10)*

C. Ribs—12 separated by intercostal spaces

1. *Classifications*

i. "true ribs"—ribs 1–7—articulate with the sternum and vertebrae

a. have tubercles (vertebrosternal) (vertebrocostal)

b. Superiorly to inferiorly they get larger.

ii. "false ribs"—ribs 8–12

a. 8–12 combine with cartilage which articulates with the sternum.

b. 11 and 12 are sometimes called "floating ribs"—they have a single facet and no neck or tubercle

c. Superiorly to inferiorly the "false ribs" become shorter.

iii. "typical ribs"—ribs 3–10

a. portions

1) head—posterior—has facets—articulates with the vertebral body of its own number and that of the vertebra below. The head of a rib and a vertebral body are connected by the stellate (radiate) ligament.

2) neck—flat—between the head and tubercle

3) tubercle—a surface projection which articulates with the transverse process. The tubercular (costotransverse) ligament extends between a rib and a vertebral transverse process.

4) body (shaft)—the upper border is curved. The lower border is grooved for the vein (located superiorly), the artery (located in the middle), and the nerve (located inferiorly).

2. *Rib articulations—ribs 2–10 articulate with the sternum via synovial joints.*

Inflammation can produce costochondritis, sometimes called "Tietze's Syndrome."

3. *Individual ribs*

i. cervical rib—from C7 (or T1)

ii. rib 1

a. is short and flat from above downward

b. has the scalene tubercle on the inner border

What the surgeons call the thoracic outlet anatomists call the superior aperture or inlet.

III. THORACIC "CAGE" (170, 176, 177)

A. Boundaries

1. *Superior aperture (also called the inlet)—manubrium, 1st rib, the intervertebral disk at C7-T1*

2. *Inferior aperture (also called the anatomical outlet)—T12, xiphosternal joint, diaphragm*

3. *Anterior—sternum*

4. *Posterior—thoracic vertebrae*

B. Articulations

1. *Costovertebral articulation*

i. vertebrae with heads of ribs

ii. tubercles with transverse processes (except 11 and 12)

2. *Sternocostal articulation*

i. manubrium with 1st rib (usually fused)—lifts on deep respiration

ii. sternum with costal cartilages of ribs 2 to 7—causes elastic recoil

iii. xiphoid with sternal body—a cartilaginous articulation reinforced by ligaments

IV. PECTORAL (SHOULDER) GIRDLE (170)—the junction of the scapula and clavicle to the trunk (at the sternoclavicular joint). The clavicle articulates with the manubrium and scapula.

V. SURFACE FEATURES (171)

A. Jugular notch—part of the manubrium

B. Sternoclavicular joint

C. Sternal angle—at the junction of the manubrium and sternum

D. Xiphoid process—below the sternum

VI. PECTORAL MUSCLES (174)

A. Pectoralis Major—extends from the clavicle, sternum, and costal cartilages to the intertubercular (bicipital) groove of humerus—elevates ribs 1–6—innervated by the anterior thoracic nerve

B. Pectoralis Minor—extends from the upper ribs to the coracoid process of the scapula—innervated by the anterior thoracic nerve

C. Serratus Anterior—extends from the upper ribs to the medial border of the scapula—covers lateral part of thorax—lies under the pectorals—pulls the scapula forward over the chest wall—innervated by the long thoracic nerve **(which may be injured during lymph node dissection and produce "winged" scapula)**

D. Latissimus Dorsi—from the lower thoracic vertebrae and posterior iliac crest to the intertubercular groove of the humerus—adducts the humerus, extends and medially rotates the arm—innervated by the thoracodorsal nerve

Because the latissimus crosses the inferior angle of the scapula, paralysis causes "winging" of the scapula

E. Rhomboids—from the upper thoracic vertebrae to the lower inferior scapula—innervated by the dorsal scapular nerve

VII. INTERCOSTAL SPACES (175, 176) —Vessels and nerves are located at the lower borders of the ribs.

A. Muscles—internal and external intercostals and subcostals

1. *Internal intercostals—rib above to rib below (incomplete posteriorly)—lower ribs in forced expiration*
2. *Deep internal (Innermost) muscle layer*
    i. transversus—from the vertebral column to higher ribs
    ii. innermost intercostals—from the inner surface of the rib to second rib below

B. Arteries in the Intercostal Spaces

1. *Superior (Supreme) intercostals—from the costocervical trunk—supply the upper two intercostal spaces*
2. *Branches of the internal thoracic—supply anterolateral intercostal spaces and* ***anastomose*** *with the anterior branches of the posterior (aortic) intercostals*
3. *Anterior branches of the posterior intercostals (from the aorta)—supply posterior aspects of the intercostal spaces, the vertebral column, and back muscles*

C. Veins in the Intercostal Spaces

1. *Anterior intercostal—drain into the internal thoracic vein*

2. *Posterior intercostal—drain into the azygos system*

➢VIII. ARTERIES (175, 176, 179)—from the thoracic aorta

A. Posterior Intercostals—single—from the back of the aorta

1. *Anterior branch—supplies posterolateral aspect of skin, muscle, intercostal spaces, and mammary glands.*
   Anterior branches ***anastomose*** *with the anterior intercostals of the internal thoracic. T10 and T11 intercostal spaces receive blood only from the anterior branch of the posterior intercostals.*

2. *Posterior (Dorsal) branches—supply spinal cord and posterior aspects of muscle and skin*

B. Subcostals—supply the subcostal area below the last rib. (They correspond to the anterior branch of the posterior intercostals.)

C. Superior Phrenic—to part of the diaphragm

D. Subclavian

1. *Internal thoracic*

   i. anterior intercostals—paired

      a. supplies lower anterior intercostal spaces, muscles, parietal pleura, and mammary gland

      b. branches to thoracic viscera

   ii. at level of 6th intercostal space divides into:

      a. musculophrenic—supplies intercostal spaces 7–9, skin of the breast, and the pectoral muscles

      b. superior epigastric—supplies the diaphragm and upper abdominal wall

2. *Costocervical trunk*

   i. superior (supreme) intercostal—supplies 2 uppermost intercostal spaces

IX. VEINS OF THE THORACIC WALL (175, 176)

A. Posterior Intercostals—enter the azygos system which enters the superior vena cava

B. Anterior Intercostal—enters the internal thoracic which enters the subclavian

C. Left Intercostal—above the accessory (superior) hemiazygos—enter into superior intercostals which enter the brachiocephalic

X. LYMPHATICS OF THE THORACIC WALL (227)

A. Posteriorly into the posterior axillary and vertebral nodes

B. Anteriorly into the internal thoracic (parasternal) nodes

XI. NERVES (166, 179) —11 pairs (anterior rami of T1–T11)

1. *T1—forms T1 root of the brachial plexus*
2. *T1–T6—6 upper nerves in intercostal space only*
3. *T7–T11—extend to the anterior abdominal wall*
4. *T12 (Subcostal nerve)—extends to the abdominal wall*

Pain from lower thoracic vertebrae may be referred to the abdominal wall.

The intercostal nerves:

1) are motor to muscles

2) are sensory to costal parietal pleura

3) carry sympathetic fibers to vascular smooth muscle

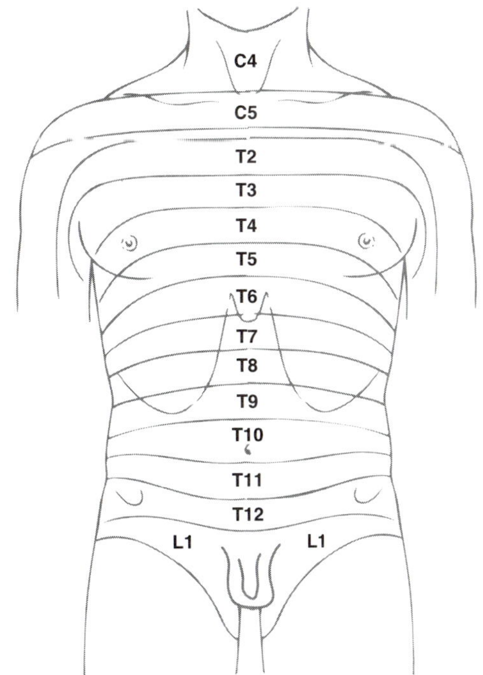

Azygos System

The azygos system is formed in the abdomen by the right ascending lumbar, renal, or IVC to pass thru the aortic hiatus to the right of the vertebrae and end in the SVC.

I. TRIBUTARIES ON THE RIGHT (226)

A. Superior (Supreme) intercostal—may drain into the brachiocephalic vein

B. Right subcostals

C. Lower 8 right posterior intercostal veins

D. Right lumbar veins

E. Bronchial veins (occasionally)

F. Hemiazygos (variable formation and drainage). It ascends to the left of the vertebrae.

II. TRIBUTARIES OF THE HEMIAZYGOS (226)—located on the left side. It arises from the left subcostal and left ascending lumbar to pass thru the aortic hiatus then behind the esophagus to enter the azygos vein.

A. Superior (Supreme) intercostal

B. Left posterior intercostal veins

C. Esophageal veins

D. Posterior mediastinal veins

E. Left lumbar veins

F. Left subcostal veins

G. Left lumbar veins

H. Accessory hemiazygos—when present may be formed by the left superior (supreme) intercostal veins. It may drain into azygos or the left brachiocephalic vein.

Breast

I. PORTIONS (167)

A. Lobes (14–20)—separated by septa to form the suspensory (Cooper's) ligament in the upper part of the breast. The glandular elements are tubuloalveolar type. Breast tissue is located in the superficial fascia.

B. Tail—can extend into the axilla

C. Areola—has sebaceous glands and smooth muscle

D. Nipple—A lactiferous duct from each lobe exits on the nipple. (Myoepithelial cells, under the influence of oxytocin, contract to express milk into the lactiferous ducts.)

II. ARTERIES (168)

A. Anterior branches of the posterior intercostals

B. Intercostal branches of the internal thoracics

C. Lateral thoracics of the axillary artery

III. VEINS (169) —drain into axillary, internal thoracic, lateral thoracic, and upper intercostal veins

IV. LYMPHATICS (169) —The glands and their ducts are surrounded by lymphatics which drain into lymphatics located between the lobes.

A. Lateral Quadrants—drain into anterior axillary (pectoral) nodes. The pectoral nodes have efferents to the apical nodes. The apical axillary nodes are located along the medial aspect of the axillary vein and behind the pectoralis minor muscle.

B. Medial Quadrants—thru intercostal spaces to the internal thoracic (parasternal) nodes. Medial quadrant lymphatics are located along the internal thoracic vessels. These lymphatics have connections to the opposite side **(important in carcinoma of the breast).**

C. Posterior Part of the Breast—to the posterior intercostal nodes

D. Upper Part of the Breast—to the apical axillary, supraclavicular, and infraclavicular nodes

V. NERVES—lateral and anterior cutaneous branches of intercostal nerves 2–6

Pleurae

I. DIVISIONS (195)

A. Parietal

1. *Costal—over the thoracic wall*
2. *Diaphragmatic—over the superior surface of the diaphragm*
3. *Mediastinal—over the lateral aspects of the mediastinum*
4. *Apical—just below the subpleural membrane of the thoracic inlet. The cupula covers the apex of the lung and lies at the level of the 1st rib.*

B. Visceral—closely applied to the lungs—lacks pain fibers

II. REFLECTIONS

A. Pulmonary Ligament—the area where the visceral and parietal layers become continuous and enclose the root of the lung. It is a fold which allows for movement of the lung during respiration.

B. Sternal Reflection—the continuation of the costal pleura with the mediastinal pleura

C. Costal Reflection—the continuation of the costal pleura with the diaphragmatic pleura

D. Vertebral Reflection—the continuation of the costal pleura and mediastinal pleura along the sides of the thoracic vertebrae

III. RECESSES (184, 185, 195)—The lung does not extend into the recesses during quiet breathing but does so during deep respiration.

A. Costodiaphragmatic—the angle between costal and diaphragmatic pleurae—marks the lower limit of the pleurae. The lung does not extend here during quiet breathing but does so during deep respiration.

B. Costomediastinal—a thin pleural space, on the left side, between the levels of the 4th and 5th costal cartilages. The area is known as the cardiac notch.

IV. NERVES (182)

A. Costal (Parietal) Pleura—supplied by sensory pleural branches of the intercostal nerves (anterior rami of spinal nerves). The costal pleura has pain fibers.

B. Mediastinal Pleura—supplied by the phrenic nerve

C. Diaphragmatic Pleura

1. *Dome—phrenic nerve*
2. *Peripherally—lower 5 intercostal nerves*

Lungs

➢ **I. RESPIRATORY TRACT (190–192)—The portion of the respiratory tract above the bronchioles consists of pseudostratified, ciliated columnar epithelium with goblet cells.**

A. Bronchi—contain C-shaped cartilages, smooth muscle, and mucoserous glands. (The left main stem bronchus is more horizontal than the right.)

B. Bronchioles—epithelium is simple columnar to cuboidal

C. Alveoli—type I cells for gas exchange—type II secrete surfactant

➢ **II. LOBES (188, 189, 191)—All lobes, with the exception of the superior lobe of the right lung, contact the diaphragm. Bronchopulmonary segments contain a segmental artery and segmental bronchus.**

A. Right Lung—larger than the left

1. *Superior (Upper)—3 segments (apical, posterior, and anterior)*
2. *Middle—2 segments (lateral and medial)*
3. *Inferior (Lower)—5 segments (superior, medial, anterior, lateral, and posterior basal)—(site of aspirated objects)*

B. Left Lung

1. *Superior—4 segments (apical, posterior, anterior superior, and inferior)*
2. *Inferior—4 segments (superior, anterior, medial lateral, and posterior basal)*

➢ **III. FISSURES (186, 187)**

A. Right Lung—The right lung has horizontal and oblique fissures.

1. *Horizontal fissure—separates superior and middle lobes. It runs at the level of the 4th costal cartilage to meet the oblique fissure at the level of the 5th or 6th rib in the midaxillary line. The horizontal fissure* ***projects*** *onto the 5th intercostal space anteriorly, 6th rib in the midaxillary line.*
2. *Oblique fissure—separates superior and inferior lobes. The oblique fissure follows a curved line from the spine of T3 to the 6th rib anteriorly. At the level of the 6th rib the oblique fissure is in the midaxillary line.*

B. Left Lung—has only an oblique fissure

1. *Oblique fissure—separates superior and inferior lobes*

➢ **IV. ROOTS (187)—The lungs are attached to the heart and bronchi by roots via pulmonary ligaments (pleura reflected off the lung onto the**

mediastinum and surrounding the pulmonary vessels). The roots are composed of bronchi, pulmonary arteries and veins, lymphatics, and nerves.

V. RELATIONSHIPS (184, 185, 195)

A. Right Lung—superior vena cava, inferior vena cava, azygos, vein, phrenic nerve, heart

B. Left Lung—aorta, esophagus, subclavian artery, heart

C. The inferior borders normally lie at the level of T10 posteriorly, the 10th rib in the midscapular line, the 8th rib in the midaxillary line, and the 6th rib in the midclavicular line.

D. Apex—extends to about 1 inch above the clavicle. The subclavian artery arches over the pleural dome.

E. Heart—displaces the left lung to form the cardiac notch

VI. ARTERIES (194–196)

A. Bronchial—from the aorta (usually one on the right, two on the left)—divide with the bronchi to supply bronchi and bronchioles

B. Pulmonary Arteries—branches follow bronchi and ultimately surround alveoli. The trunk and right and left branches lie inferior to the aortic arch. The ligamentum arteriosum (formerly ductus arteriosus) extends to the inferior portion of the arch.

VII. VEINS (194–196)

A. The Bronchial Veins—exit the root of the lungs and empty into the azygos system

B. The Pulmonary Veins—end in the left atrium. (They contain oxygenated blood.)

VIII. LYMPHATIC FLOW (197)—1) pulmonary (lobar bronchi) to 2) bronchopulmonary (hilar) to 3) tracheobronchial to 4) bronchomediastinal and parasternal to enter 5) the right lymphatic duct on the right or the thoracic duct on the left side

IX. NERVES (198, 199)—via pulmonary plexus located in the hila

A. Sympathetic (T2–T5)—produce bronchodilation and vasoconstriction

B. Parasympathetic (P)—from the Vagus nerves—cause the bronchi to constrict and the vessels to dilate. The parasympathetics go to the pulmonary plexus and follow the vessels.

C. Visceral Afferents—Pain fibers go along sympathetic pathways while stretch components accompany the Vagus.

Anatomical Basis of Respiration

I. QUIET INSPIRATION (183)

A. Diaphragm—extends from lower costal cartilages and upper lumbar vertebrae to the central tendon

B. Muscles Which Elevate the Ribs

1. *External intercostals—rib above to rib below—elevate the ribs in inspiration*
2. *Internal intercostals—rib above to rib below*
3. *Deep internal (Innermost) muscle layer—transversus, innermost intercostals, and subcostals.*

Paralysis of the intercostal musculature results in bulging of intercostal spaces on expiration.

II. FORCED INSPIRATION (183)—can produce maximum increase in the capacity of the thoracic cavity

A. Muscles Which Elevate the Ribs

1. *Scalenus anterior and medius—extend from cervical vertebrae to the upper surface of the 1st rib*
2. *Sternocleidomastoid—extend from occipital and mastoid to elevate the manubrium and clavicle*

Long-distance runners put their heads back when they reach the finish line.

B. Muscles Which Fix the Scapula—levator scapulae, trapezius, and rhomboid

C. Muscle Which Fixes the Upper Limb—pectoralis major. With the upper limb fixed, the pectoralis major can pull upward on the sternum and upper ribs and increase lung capacity.

Tired football players put their hands on their hips.

III. QUIET EXPIRATION (183)—essentially passive

IV. FORCED EXPIRATION (183)

A. Muscles Which Pull Down the Ribs (The quadratus lumborum and transverse thoracic muscles **may** also assist.)

1. *Abdominal muscles*
2. *Latissimus dorsi—from the lower thoracic vertebrae to the humerus*
3. *Internal intercostals (and possibly the internal thoracic)*

Infants have horizontal ribs and therefore use abdominal muscles in respiration. Males use more abdominal respiration than females.

Diaphragm

I. EMBRYOLOGY

A. Septum Transversum—initially lies in the cervical area.
The phrenic nerve from C3, C4, and C5 grows into the septum.
The septum forms the tendinous portion of the diaphragm.
The "Descent" of the diaphragm is due to relatively rapid growth of dorsal vertebrae and lung expansion.

B. Pleuroperitoneal Membranes (2)

1. *Separate thoracic and abdominal portions of the coelom*
2. *If membranes do not completely form over openings (canals) or the diaphragm is incompletely formed, a congenital diaphragmatic hernia may result.*

Incomplete closure of the pleuroperitoneal canal of Bochdalek located at the lumbocostal (vertebrocostal) triangle may result in a diaphragmatic hernia.

Incomplete closure of the sternocostal triangle (lower parasternal) (the foramen of Morgagni) may result in a congenital diaphragmatic hernia.

C. Myoblasts from body walls form muscular components of the diaphragm.

D. Dorsal Mesentery of the Esophagus (a portion of dorsal mesentery of the gut) provides a pathway for the vessels, nerves, and myoblasts (which form the crura of the diaphragm).

II. STRUCTURE (181)—The diaphragm is dome shaped. The domes of the diaphragm lie at the level of the 5th rib.

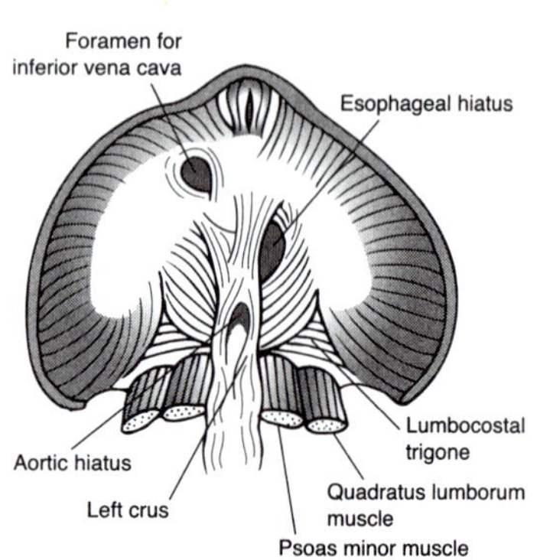

A. Muscular Origins—extend centrally to insert in the central tendon

1. *Sternal*
2. *Costal—lower 6 ribs*
3. *Vertebral (Lumbar)—crura originate from lumbar vertebrae (L1–L3)*

B. Fascial Origins of the Diaphragm

1. *Medial arcuate ligament (Medial lumbocostal arch) (Internal arcuate ligament)—extends over the surface of the psoas to L1*
2. *Lateral arcuate ligament (Lateral lumbocostal arch) (External arcuate ligament)—extends from L1 vertebra to run along the 12th rib—goes over the quadratus lumborum*

III. OPENINGS (FORAMINA, APERTURES, HIATI) (181)

A. Aortic Hiatus—located between the crura.
It transmits the aorta, azygos, and thoracic duct.

B. Esophageal Hiatus—located at the level of T10. It transmits the esophagus, vagi, branches of gastric vessels, and the lymphatics of the lower third of the esophagus **and is the site of hiatal hernias.**

C. Caval Foramen—located in the central tendon at the level of T8. It transmits the inferior vena cava and right phrenic nerve.

IV. ACTIONS OF THE DIAPHRAGM

A. Muscle of Inspiration—When it moves downward, intrathoracic pressure is decreased.

B. Muscle of Abdominal Straining—By raising intra-abdominal pressure, the diaphragm aids in evacuation of pelvic contents and the return of venous blood to the right atrium.

C. Muscle of Weight Lifters—Fixation of the diaphragm helps support the vertebral column.

➢ V. ARTERIES (181)

A. Internal Thoracic—anteriorly

B. Intercostals (from the aorta)—to the posterior diaphragm

C. Inferior Phrenic (from the aorta)—to the inferior diaphragm

VI. VEINS

A. Superior Surface—to internal thoracic veins

B. Inferior Surface—to inferior phrenic veins

1. *Right inferior phrenic to the IVC*
2. *Left inferior phrenic to the left suprarenal vein*

➢ VII. LYMPHATICS (227)

A. Anterior diaphragm—to retrosternal and internal thoracic nodes

B. Posterior diaphragm—to mediastinal and intercostal nodes

➢ VIII. PHRENIC NERVE (176, 182)—via the cervical plexus

A. The phrenic nerve (C3–C5) is the motor and sensory nerve of the diaphragm.

B. It lies medial to the internal jugular in the neck, then enters the thorax posterior to the subclavian vein to descend along the pericardium and enter the diaphragm.

C. Pain from the diaphragm may be referred to C2, C3, and C5 dermatomes, and therefore to the shoulder and the lateral aspects of the neck.

Trachea

The trachea is approximately 5 inches long and 1 inch in diameter. It is kept open by cartilaginous rings which form a C. It is located in the superior mediastinum. The trachea begins in the neck below the larynx (cricoid cartilage) at the level of C6 posteriorly and runs down and slightly backward. It ends at the level of the sternal angle (T4) at the carina (a ridge inside the trachea) by dividing into main stem bronchi.

I. DIVISIONS (190)

A. Right Bronchus

1. *Wider, shorter, and more vertical than the left*
2. *Divides into 3 segmental bronchi going to the 3 lobes of the right lung*
3. *The azygos vein goes over the pulmonary vein inferior and anterior to the right bronchus.*

B. Left Bronchus

1. *Narrower, longer, and more horizontal than the right*
2. *Divides into 2 segmental bronchi going to the 2 lobes of the left lung*
3. *Extends under the aortic arch*

II. RELATIONSHIPS (195)—in the superior mediastinum

A. Anterior—thymus, left common carotid artery, arch of the aorta

B. Posterior—esophagus, left recurrent laryngeal nerve, thoracic duct

C. Right—azygos vein, Vagus nerve, lung, right phrenic nerve

D. Left—arch of aorta, left common carotid artery, subclavian artery, left phrenic nerve, lung, Vagus nerve

III. ARTERIES (196)

A. In the neck—branches of inferior thyroid artery

B. In the thorax—branches of descending aorta

IV. VEINS—the azygos vein and the inferior thyroid vein (which goes over the right bronchus)

V. LYMPHATICS (197, 227)—to tracheobronchial and bronchopulmonary nodes and thence to internal thoracics and bronchomediastinals

VI. NERVES (198, 199)—Vagus, laryngeals, and sympathetics

Esophagus

I. GENERAL—The esophagus (221) begins at the lower margin of the cricopharyngeus. It extends from the lower pharynx (at the level of C6 vertebra) to the stomach (at the level of T10).

A. Length—about 10 inches

B. Muscularis Externa—striated in the cervical area, smooth in the thoracic area (lower ⅔)

C. Lined by squamous epithelium

II. RELATIONSHIPS (221, 227)

A. In the Neck

1. *Anterior—trachea and recurrent laryngeal nerve*
2. *Lateral—thyroid*
3. *Posterior—vertebrae*

B. In the Thorax

1. *Anterior—trachea, left recurrent laryngeal nerve, left atrium*
2. *Posterior—thoracic vertebrae, thoracic duct, right Vagus*
3. *Right posterior—azygos*
4. *Left—left subclavian, arch of the aorta*

C. In the Abdomen

1. *Anterior—liver*
2. *Posterior—the left crus of the diaphragm*

III. ARTERIES, VEINS, LYMPH NODES

TABLE 4-1.

	UPPER ESOPHAGUS	MIDDLE ESOPHAGUS	ABDOMINAL ESOPHAGUS
Arteries (225)	Inferior thyroid—from the thyrocervical trunk	Descending aorta	Left gastric—from the celiac
Veins (226)	Inferior thyroid	Azygos system	Left gastric (Portal System)
Lymph Nodes (227)	Cervical	Posterior mediastinal	Nodes located along the left gastric and celiac arteries

IV. NERVES (228, 229)

A. Parasympathetic—via the Vagus nerve

1. *The esophageal plexus (formed by both vagal nerves) lies on the left anterior surface and the vagal trunk lies on the right posterior surface of the esophagus.*
2. *Vagal fibers innervate the longitudinal muscle layer and shorten the esophagus.*
3. *The upper esophagus is innervated via the recurrent laryngeal (of the Vagus) while the lower esophagus is innervated via the pulmonary plexus.*

B. Sympathetic—via the sympathetic trunk (cervical and thoracic chains)

1. *The superior cervical ganglion connects with the ganglion of the Vagus.*
2. *In the thorax, sympathetic fibers proceed from the upper 4 thoracic ganglia and the greater and lesser splanchnic nerves.*
3. *In the abdomen, a few fibers to the esophagus come from the celiac plexus.*

C. Visceral Afferents—to T1–T5 via thoracic splanchnic nerves

V. PERISTALSIS

A. Primary—initiated by swallowing

B. Secondary—does not require swallowing movement

VI. SPHINCTERS (222, 223)

A. Upper (Cricopharyngeal) Sphincter—Upon swallowing, it relaxes to allow food to pass; then it contracts.

B. Lower Esophageal Sphincter (LES)—A high-pressure zone in the lower 4 cm of the esophagus relaxes to allow passage of food into the stomach. Relaxation of the LES is related to ENS inibitory neurons employing VIP and NO transmitters.

VII. CONSTRICTIONS (NARROWINGS) (221)—caused by the arch of the aorta, the left main bronchus, and the diaphragm. There is also slight constriction 1 inch above the diaphragm.

VIII. ESOPHAGEAL ABNORMALITIES

A. Diverticula—may be lined by one or more layers of the esophageal wall

1. *Zenker's—located above the upper esophageal sphincter. Mucosa passes thru the cricopharyngeus muscle.*
2. *Near the midpoint of the esophagus—frequently asymptomatic*
3. *Just above the lower esophageal sphincter (LES)—may be related to motor abnormalities*

B. *Gastroesophageal Reflux—Reflux of gastric contents into the esophagus may cause symptoms of heartburn, esophagitis, strictures, ulceration, or metaplasia.*

C. Barrett's Syndrome (esophagitis, weight loss, and dysplasia)—replacement of squamous epithelium with columnar epithelium

D. Boerhaave's Syndrome (chest pain, tachypnea, and tachycardia)—rupture of the esophagus due to vomiting

E. Scleroderma (a connective tissue disorder)—may affect the muscularis and submucosa, often resulting in dysphagia

F. Atresia (Congenital)—due to incomplete separation of the esophagus and trachea. The upper portion of the esophagus may dead end. Atresia may be accompanied by a tracheoesophageal fistula.

Mediastina

I. SUPERIOR MEDIASTINUM (218)

A. Boundaries (General)

1. *Inferior—a line between the disk between the 4th and 5th vertebrae and the sternal angle (junction of the body and manubrium)*
2. *Superior—top of manubrium to body of 1st thoracic vertebra. The upper limit of the superior mediastinum is the thoracic aperture.*
3. *Anterior—manubrium*
4. *Posterior—T1–T4 vertebrae*

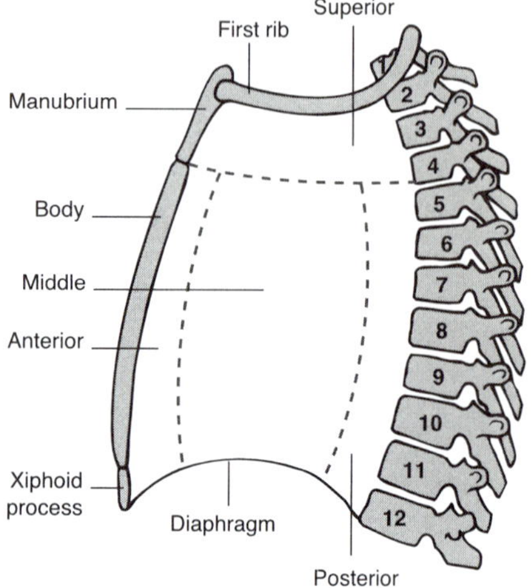

B. Contents

1. *Esophagus—lies posterior to the trachea*
2. *Trachea—is posterior to the aorta and brachiocephalic vein*
3. *Thymus*
4. *Aortic arch*
5. *SVC—lies to the right of the aorta*
6. *Brachiocephalic veins—are anterior to the aorta*

II. ANTERIOR MEDIASTINUM (218)

A. Boundaries—lies between the sternum and the pericardial sac

B. Contents—thymus (during infancy)

III. MIDDLE MEDIASTINUM (218, 230)—consists primarily of the pericardium and its contents

A. Boundaries—lies below the superior mediastinum and above the diaphragm—laterally are the lungs

B. Contents—pericardium, heart, part of the aorta, and IVC. (The phrenic nerves are on the sides of the pericardium.)

IV. POSTERIOR MEDIASTINUM (225, 227, 230)

A. Boundaries—lies posterior to the pericardium, anterior to the thoracic vertebrae (T5–T12), superior to the diaphragm, and inferior to the superior mediastinum.

B. Contents

1. *Descending aorta—on the left side*
2. *Thoracic duct—anterior to the vertebrae*
3. *Azygos veins—on the right posterior side*
4. *Hemiazygos veins—posterior left side. These veins pass behind the esophagus, thoracic duct, and aorta to enter the azygos.*
5. *Accessory hemiazygos—located on the left, where it may join the hemiazygos*
6. *Esophagus—begins on the right of the aorta and then crosses over to the left*
7. *Sympathetic trunks*
8. *Vagus*
    - i. right Vagus—located along the trachea
    - ii. left Vagus—located between the aorta and pulmonary artery, posterior to the root of the lung, and then down along the esophagus

V. OTHER CLASSIFICATIONS OF MEDIASTINA

A. Anterior and Posterior

1. *Anterior*
    - i. superior—includes the great vessels and thymus
    - ii. inferior—includes the pericardium and heart
2. *Posterior—includes trachea, bronchi, esophagus, aorta, and thoracic duct*

B. Superior and Inferior

1. *Superior*
2. *Inferior*
    - i. anterior
    - ii. middle
    - iii. posterior

Pericardium

The pericardium is a sac which encloses the heart and the roots of the great vessels. Parietal and visceral portions of the serous layer of pericardium become continuous around great vessels and form "lines of reflection."

I. LAYERS (200, 201, 203)

A. Fibrous—is thick, limits the movement of the heart, and is attached to the central diaphragm tendon

B. Serous—The pericardial sac (a potential space) lies between the parietal and visceral layers.

When the pericardial sac is opened, a cardiac surgeon can place a finger between the ascending aorta and the pulmonary veins.

Blood or pus in the pericardial sac may compress the pulmonary veins and heart and result in cardiac tamponade.

1. *Parietal layer—lines the fibrous pericardium—reflected around roots of great vessels and is continuous with epicardium (visceral serous layer). It is "pierced" by 8 vessels. Pain fibers accompany the intercostal nerves.*

2. *Visceral layer—Epicardium*

Pericardiocentesis may be performed between the left 5th and 6th intercostal spaces.

II. RELATIONSHIPS OF THE PERICARDIAL SAC (203)

A. Anterior—internal thoracic vessels and remains of the thymus

B. Lateral—parietal pleura and phrenic nerves

C. Posterior—esophagus and aorta

D. Inferior—diaphragm

III. SINUSES (203)

A. Transverse Sinus—located at the convergence of the visceral and parietal pericardia—connects the 2 sides of the pericardial cavity

1. *Relationships*

i. Posterior to the sinus lie the superior vena cava, left atrium, and superior left pulmonary vein.

ii. Anterior to the sinus lie the aorta and the pulmonary arteries.

B. Oblique Sinus—a "potential space" behind the heart—the reflection of the serous pericardium around the large veins. The oblique sinus lies within the pericardial sac.

Heart

I. EXTERNAL FEATURES (201, 202, 208)

A. Surfaces

1. *Anterior*
   - i. right ventricle—contributes to most of the anterior surface
   - ii. pulmonary trunk
   - iii. right auricle
   - iv. superior vena cava
   - v. aorta
2. *Posterior*
   - i. right atrium—forms the right border of the heart
   - ii. left atrium—forms most of the base of the heart
   - iii. left ventricle
3. *Inferior—mainly the left ventricle*

B. Grooves

1. *AV sulcus (Coronary groove)—the surface separation of the atria and ventricles*
   - i. the right side contains:
     - a. the right coronary artery
     - b. the small cardiac vein
   - ii. the left side contains:
     - a. the circumflex branch of the left coronary artery
     - b. the coronary venous sinus
2. *Interventricular sulcus (groove)—the surface separation of the ventricles*
   - i. the anterior interventricular sulcus contains:
     - a. a branch of the left coronary artery
     - b. the great cardiac vein
   - ii. the posterior interventricular sulcus contains:
     - a. a branch of the right coronary artery
     - b. the middle cardiac vein

3. *Sulcus terminalis—the external separation of the auricle and main cavity of the atrium. (On the inside of the heart it corresponds to ridges called crista terminalis.)*

C. Borders

1. *Right—3rd costal cartilage to 6th costal cartilage*
2. *Inferior—6th costal cartilage (xiphisternal joint) to the apex. (The apex is formed by the left ventricle and is located superficially at the 5th intercostal space.)*
3. *Left border—apex to second left intercostal space*
4. *Superior—second left costal cartilage or intercostal space to the 3rd right costal cartilage*

D. The chambers form portions of the following:

1. *Right atrium → right border*
2. *Right ventricle → anterior surface*
3. *Left atrium → base*
4. *Left ventricle → apex*

II. MYOCARDIUM (209)

A. Covered by Visceral Pericardium (Epicardium)

B. Lined by Endothelium (Endocardium)

III. CORONARY ARTERIES (204–207)

A. Right Coronary Artery—originates in the aortic sinus and is located in the right AV sulcus (coronary groove). The right coronary artery supplies oxygenated blood to the right atrium, right ventricle, the septum, and the nodes. Its branches have **anastomoses** with the circumflex branch of the left coronary and the anterior interventricular (descending) branch of the left coronary artery.

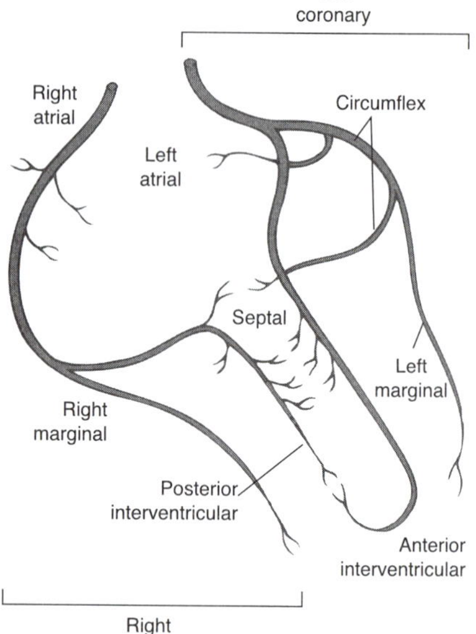

1. *Marginal—located near the inferior border of the heart—accompanies the small cardiac vein—supplies the right ventricle*
2. *Posterior interventricular (Posterior descending) branch—the continuation of the right coronary on the back of the heart. It is located in the posterior interventricular sulcus, accompanies the middle cardiac vein, and supplies the posterior wall of the right ventricle and the interventricular septum.*

B. Left Coronary Artery—originates in the left aortic sinus and is located in the AV sulcus (coronary groove). It supplies the left atrium, left ventricle, and septum. Its branches accompany the great cardiac vein. They have anastomoses with the posterior branch of the right coronary artery and the anterior branch of the right coronary artery.

1. *Anterior interventricular (Descending) artery—in the anterior interventricular sulcus—supplies the anterior wall of the ventricles and the anterior interventricular septum. It has* ***anastomoses*** *with the posterior interventricular artery (of the right coronary).*

2. *Circumflex artery—located in the posterior AV (coronary) sulcus to supply the left posterior heart (atrium and ventricle)*

   i. marginal branch

IV. CARDIAC VEINS (203–205)

A. Coronary Sinus—drains into the right atrium below the location of the AV node—has a valve

1. *Great cardiac vein—begins at the apex of the heart and lies in the interventricular groove, where it follows the anterior interventricular (descending artery) of the left coronary. (The great cardiac vein contributes to drainage of all chambers except the right atrium.)*

2. *Middle cardiac vein—begins at the apex and lies in the posterior interventricular groove—follows the posterior interventricular (descending) of the right coronary artery*

B. The Anterior Cardiac Veins—usually drain directly into the right atrium

C. Small Cardiac Vein—follows the marginal branch of the right coronary artery and drains into the right atrium

V. INTERNAL FEATURES (208–212)

A. Right Atrium (208)

1. *Entering vessels—The sinus venosum receives the venae cavae (SVC and IVC) and the coronary sinus.*

2. *Features*

   i. crista terminalis—separates the main cavity and the auricle

   ii. pectinate muscles—in the walls of the auricle. The pectinate muscles extend from the crista.

   iii. fossa ovalis—a depression in the interatrial wall (atrial septum)—a remnant of the former foramen ovale

   iv. upper surfaces of cusps of tricuspid valve

➢ B. Left Atrium (209)

1. Entering vessels

i. pulmonary veins (4)—**no** valves

ii. coronary sinus—receives the cardiac veins

2. Features

i. pectinate ridges—in the auricle

ii. upper surfaces of cusps of the mitral valve

➢ C. Right Ventricle (208–211)

1. Exiting vessel—pulmonary artery

i. has 3 semilunar cusps which form the pulmonary valve

a. nodule—a central thickening on the edge of each cusp

b. lunulae—small areas located on the sides of the nodule

c. sinus—the pocket behind the cusp

ii. The conus arteriosus (infundibulum) is the upward extension of the right ventricle into the pulmonary trunk.

2. Features

i. tricuspid valve (3 cusps)—the AV valve

a. anterior

b. posterior (or right)

c. septal (or left)

ii. cordae tendinae—connect cusps to the papillary muscle

iii. the wall of the left ventricle is thicker than the wall of the right ventricle

iv. trabeculae—muscular ridges

v. moderator band—trabeculae which extend from the septum to the anterior papillary muscle (involved in conduction)—contains the AV bundle of His

➢ D. Left Ventricle (209)

1. Exiting vessel—the infundibulum of the aorta and its valve (semilunar with 3 cusps—similar to those of the pulmonary artery)

i. The right aortic sinus gives origin to the right coronary artery.

ii. The left aortic sinus gives origin to left coronary artery.

2. *Features*

i. mitral valve (bicuspid)—has 2 cusps—**the valve commonly subject to disease**

ii. cordae tendinae—connect cusps to the papillary muscle—The papillary muscle and cordae maintain the position of the valve leaflets.

iii. trabeculae—muscular ridges which form the inflo tract

iv. wall of left ventricle thicker than right ventricle

v. moderator band—contains the AV bundle of His

E. General Features of AV Valvular Closure (210–212)

1. *Annulus (fibrosis)—a ring of connective tissue around the valves*

i. provides attachment for cardiac muscle

ii. separates cardiac atrial and ventricular musculature

2. *Cusps—thin connective tissue leaflets attached to the annulus*

3. *Cordae—slender cords of connective tissue connecting cusps with papillary muscle*

4. *Papillary muscles—stout pillars of cardiac muscle which contract first and bring about closure of the valves*

VI. CONDUCTING SYSTEM (213)

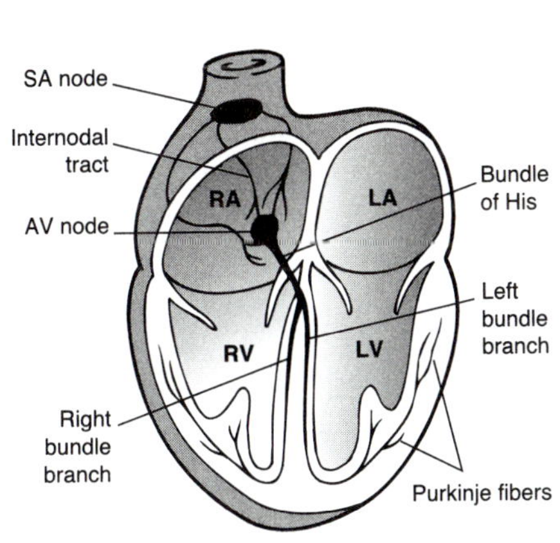

A. Sinoatrial Node (SA node) (Pacemaker)

1. *Located to the right of the SVC in the right atrium*

2. *The atrial (nodal) branch of the right coronary artery supplies the SA node.*

3. *Initiates cardiac contraction by muscle depolarization*

4. *When influenced by sympathetic fibers, the result is cardiac acceleration.*

5. *When influenced by parasympathetic fibers (Vagus), the result is cardiac slowing.*

6. *When it fails, the AV junction takes over.*

B. Atrioventricular Node (AV node)

1. *Located in the atrial septum beneath the endocardium*
2. *The right coronary artery supplies the AV node.*
3. *Changes in the AV node cause changes in the P-R interval on the EKG.*
4. *A block of the AV junction can cause separate actions of the atria and ventricles.*

C. AV Bundle of His

1. *Passes thru the annulus fibrosis*
2. *Is located in the interventricular septum (from the membranous portion to the muscular portion)*
3. *Divides into right and left branches, one to each ventricle*

D. Purkinje Fibers—a subendocardial plexus of specialized cardiac muscle fibers that form the conducting system.

NOTE: The autonomic nervous system can influence the conducting system (parasympathetics slow and sympathetics accelerate). The moderator band contains conducting fibers. It extends from the septum to the anterior papillary muscle.

VII. CARDIAC PAIN (214, 215)

A. Visceral Afferents travel up along the cardiac nerves and splanchnic nerves in a retrograde fashion to enter the sympathetic chain in the cervical region and thence **down** the chain to enter spinal nerves via white rami at the levels of T1–T4 on the left.

B. The above forms the basis for referred cardiac pain.

VIII. VALVES

TABLE 4-2.

VALVE	LOCATION	AREA OF AUSCULTATION
Tricuspid	Right 4th, 5th, and 6th costal cartilages	6th interspace to the right of the sternum and over the right half of the inferior end of the body of the sternum
Mitral	Left side of the sternum at the level of the 4th costal cartilage	Apex of the heart (left 5th intercostal space) in approximately the midclavicular line
Pulmonary	Left 3rd costal cartilage	2nd left intercostal space
Aortic	Left sternal border at the level of the 3rd costal cartilage	2nd right intercostal space at the edge of the sternum

IX. CARDIAC ABNORMALITIES

A. Septal defects—may create left-to-right shunts

1. *Atrial septal defect (ASD)—located at the site of the former foramen ovale and may result in pulmonary hypertension*

2. *Ventricular septal defect (VSD)*

B. Patent Ductus—The ductus venosus remains open at birth, resulting in flow from the aorta into the pulmonary artery.

C. Aortic Valve Stenosis—congenital or acquired—results in left ventricular hypertrophy (LVH) and congestive heart failure

D. Coarctation of the aorta—may occur above or below the ligamentum artertiosum

E. Tetralogy of Fallot—A congenital abnormality resulting in cyanosis in the newborn. It is the result of a misplaced septum.

1. *Ventricular septal defect—producing a right-to-left shunt*

2. *An aorta which "overrides" the ventricles*

3. *Pulmonary stenosis—producing RVH*

4. *Right ventricular hypertrophy (RVH)*

F. Cardiac Tamponade—Fluid in the cardiac sac produces pressure on the pulmonary veins, causing increased venous pressure.

Chapter

Abdomen

Page numbers in black. Netter plate numbers in green.

Abdominal Aorta (247)

I. INFERIOR PHRENIC

A. Superior Suprarenal

II. LUMBAR (4)

III. CELIAC

A. Left Gastric

B. Splenic

C. Common Hepatic

IV. MIDDLE SUPRARENAL

V. SUPERIOR MESENTERIC

A. Inferior Pancreaticoduodenal

B. Middle Colic

C. Right Colic

D. Ileocolic

E. Jejunal and Ileal

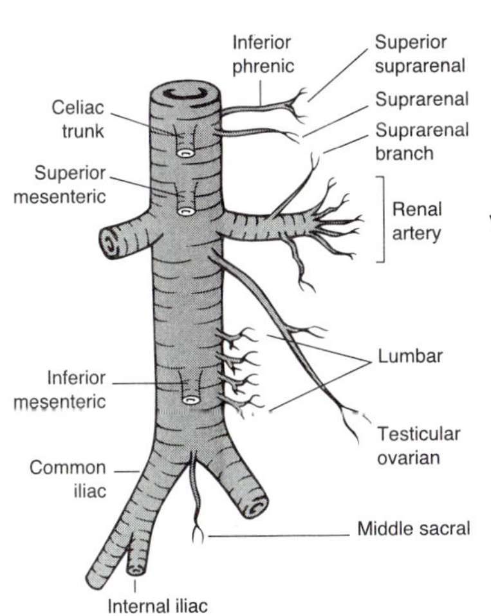

VI. RENAL

A. Inferior Suprarenal

VII. TESTICULAR/OVARIAN

VIII. INFERIOR MESENTERIC

A. Left Colic

B. Sigmoid (3)

C. Superior Rectal

IX. MIDDLE SACRAL

X. COMMON ILIACS

XI. LEVELS OF SELECT BRANCHES OF THE AORTA

A. Celiac—T12

B. Superior mesenteric—L1

C. Renal L2

D. Inferior mesenteric—L3

E. Common iliacs—L4

Inferior Vena Cava

The inferior vena cava (IVC) (248) receives tributaries from the lower limbs, the abdominal wall, and (via the portal system) the abdominal viscera. The IVC is formed by common iliacs. It proceeds into the middle mediastinum thru the caval opening along with the phrenic nerve.

I. RIGHT AND LEFT HEPATICS—the end of the portal system

II. INFERIOR PHRENICS—drain the diaphragm

III. RIGHT SUPRARENAL—the left drains into the left renal

IV. LUMBAR VEINS—4 on each side

V. AZYGOS—alternative connection between SVC and IVC

VI. RENAL—the left receives blood from the left suprarenal

VII. TESTICULAR/OVARIAN—the left drains into the left renal

VIII. COMMON ILIACS

  A. Internal iliacs—drain the pelvis

  B. External iliacs—drain the lower limbs

IX. MEDIAN SACRAL (may enter the left common iliac vein)

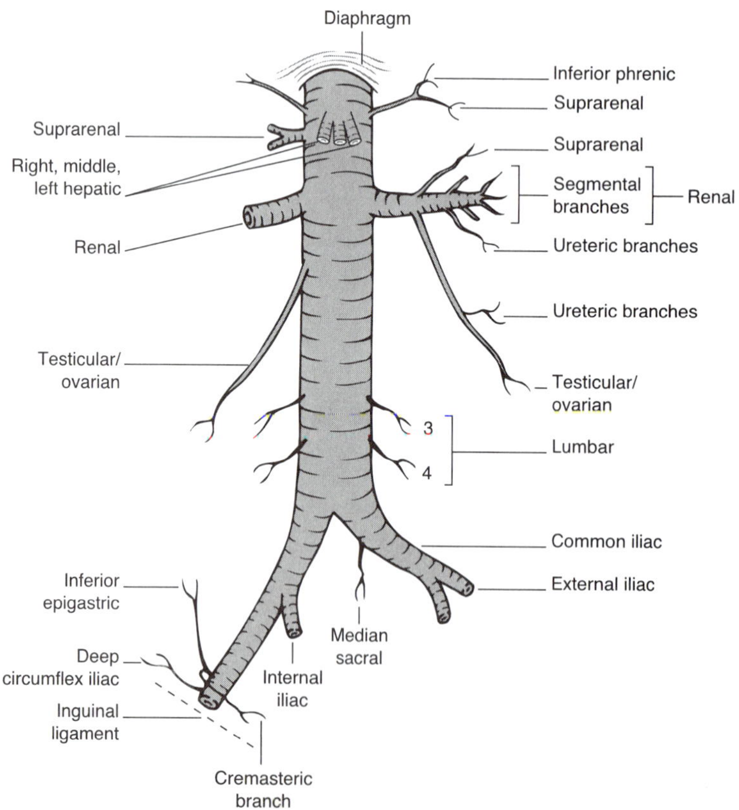

Nervous Systems in the Abdomen

I. FIBERS OF THE SOMATIC NERVOUS SYSTEM (250)

A. Intercostal and Lumbar Nerves—supply the anterior abdominal wall skin, muscles, and peritoneum

B. Lumbar Plexus (L1–L4)

1. *Formed in the psoas from anterior rami*
2. *Also receives fibers from the sympathetic trunk*

II. FIBERS OF THE ANS (306)

A. Sympathetic (SNS)

1. *Synapses usually take place in paravertebral (sympathetic chain) or prevertebral ganglia and travel via splanchnic nerves. Foregut derivatives synapse in celiac ganglia.*
2. *Major organs supplied:*
   - i. stomach—SNS inhibits juices, decreases motility, and contracts the pyloric sphincter
   - ii. pancreas—SNS decreases enzyme secretion
   - iii. spleen—SNS contracts
   - iv. small intestine—SNS decreases motility
   - v. proximal ⅔ of colon—SNS decreases motility
   - vi. kidney—SNS constricts vessels
   - vii. suprarenal—SNS increases catacholamines

B. Parasympathetic (PNS)

1. *Parasympathetic fibers pass thru preaortic ganglia without a synapse. In the abdomen they are usually contained in the Vagus and terminate in cell bodies close to the organ.*
2. *Major organs supplied:*
   - i. stomach—PNS stimulates gastric juices
   - ii. liver—PNS stimulates bile and gluconeogenesis
   - iii. pancreas—PNS increases secretions
   - iv. intestines—PNS increases peristalsis
   - v. upper ureters—PNS increases peristalsis
3. *Pelvic splanchnics supply:*
   - i. distal ⅓ of colon—PNS increases peristalsis
   - ii. sigmoid colon and rectum—PNS increases peristalsis

Visceral Afferent Nerves in the Abdomen

Visceral afferents share pathways with the Autonomic Nervous System (ANS). Because pain fibers run along sympathetic pathways, sympathectomy may be employed to relieve visceral pain.

I. PATHWAYS (306)

A. Shared Parasympathetic Pathways—carry reflex afferents along with Vagus fibers to the brain (e.g., gastric reflex)—stretch or noxious stimuli in the GI tract may initiate transmitter release (e.g., Substance P) and thus ENS communication along afferent pathways to dorsal root ganglia, the Vagal (Nodose) ganglia, and the CNS. Neurotransmitter release may also activate spinal reflexes.

B. Shared Sympathetic Pathways—carry pain and pressure afferents via thoracic and lumbar splanchnics

1. *Foregut derivatives (esophagus, stomach, gallbladder, and 1st and 2nd parts of the duodenum) travel via the celiac plexus and greater splanchnic nerve to vertebral levels T5–T9; therefore, pain can be referred to the lower thoracic and epigastric regions.*
2. *Midgut derivatives (2nd and 3rd parts of the duodenum, jejunum, ileum, appendix, ascending colon, and transverse colon) travel via the superior mesenteric plexus and lesser splanchnic nerves to vertebral levels T10 and T11; therefore, pain can be referred to the umbilical region.*
3. *Hindgut derivatives (descending and sigmoid colon) and the mid-portion of the ureter travel via the aortic plexus and lumbar splanchnics to vertebral levels L1 and L2; therefore, pain can be referred to the pubic area, inguinal regions, scrotum, or labia.*
4. *Pain from the kidneys, upper ureters, and gonads travels to the aortic plexus, renal plexus, lowest (least) splanchnic nerve (T12), and lumbar splanchnic nerves (L1); therefore, pain can be referred to the lumbar and inguinal regions.*

II. REFERRED PAIN

A. **The pathways above explain the basis for referred pain.**

B. **Visceral pain from any thoracic or lumbar level is difficult to distinguish from that of somatic origin.**

C. **Pain is referred to a somatic dermatome associated with the particular splanchnic nerve.**

Enteric Nervous System

The enteric nervous system consists of a neural network and supporting cells involving the pancreas, the gallbladder, and the GI tract. Although the ENS is sometimes considered part of the ANS, it is capable of independent function. The ENS regulates secretion (endocrine and exocrine) and blood flow along the GI tract. It also regulates motility, primarily via the Migratory Motor Complex (MMC).

I. SECRETORS—The ENS causes the release of gastrointestinal peptides. These peptides can be categorized according to their structure or activity. They can be released locally or can pass thru the bloodstream. Four methods of peptide delivery are:

A. Autocrine—The cell effects its own secretion (i.e., can act on the tissue that produced it).

B. Paracrine—local diffusion of substances to target cells (i.e., can act on the tissue within the same gland)

C. Neurocrine—Substance is released by neurons innervating the target cells (i.e., can be released by nerves and act on a tissue which contains the appropriate receptors).

D. Endocrine—secretion by the cell directly into the bloodstream (and can exert effects on a tissue that has the appropriate receptors)

II. TARGET CELLS—may receive the peptide secretions above in one or more of the four ways described above. The receptor in the target cell may be activated or inhibited by the secretion.

III. MMC—originates in a gastric pacemaker located on the greater curvature of the stomach. It initiates a wave of propulsion which proceeds down the GI tract at intervals. Motilin may be involved in the MMC initiation. The myenteric and submucosal plexuses are both responsive to the ANS and ENS.

A. Peristalsis (contraction followed by relaxation). The MMC controls the peristaltic reflex which regulates smooth muscle contraction and relaxation. This action moves food particles, sheds cells and bacteria, and prevents bacterial overgrowth. Reflexes may occur entirely within the wall of the GI tract.

B. Sphincter Function—Relaxation and contraction of sphincters is under general control of the ENS.

1. *Contraction—mediated by cholinergic and adrenergic neurons*

2. *Relaxation—related to Vasoactive Intestinal Peptide (VIP), Motilin, Nitric Oxide (NO), and other mediators.*

TABLE 5-1. ENS

GLANDS	MODE OF PEPTIDE DELIVERY	EXAMPLES
Paracrine	Substances are diffused locally thru interstitial fluid to target cells	Somatostatin Nitric Oxide (NO)
Endocrine	Secretion directly into bloodstream	Gastrin Cholecystokinin (CCK)
Neurocrine	Release of substance by neurons innervating target cells	Gastrin Releasing Peptide (GRP) Substance P
Autocrine	Effects its own secretion	Somatostatin inhibits its cell of origin NO

STIMULUS	SUBSTANCES RELEASED	EFFECT
Sight, Smell	Peptides Gastrin	Vagal efferents
Food in the Gut	GRP	Release of Gastrin from antral G cells
Gastrin	Hydrogen ions from parietal cells	Increase in gastric acid
Food in the Gut	Neurohumoral agents	Gastric pacemaker begins Migratory Motor Complex (MMC) to contract and relax intestinal circular muscles
Food in the Gut	CCK	Regular emptying of chyme into the duodenum
Fat	Duodenal CCK Colonic Neurotensin Colonic Glucagon	Gallbladder contraction Vasodilation Mucosal enlargement
Intraluminal pH	Somatostatin (from mucosal and pancreatic cells)	Inhibits hormones and release of GI peptides
Intestinal Distension	Vasoactive Intestinal Peptide (VIP) NO Substance P	Smooth muscle relaxation Relaxes sphincters Contracts smooth muscle
Intraluminal Antigen plus a Sensitized Intestine	Paracrine agents	Elimination of the antigen by gut immunocytes

Anterior Abdominal Wall

I. FASCIA (232)

A. Superficial

1. *Superficial layer (Camper's) (Fatty layer)—extends into the scrotum as the dartos layer*

2. *Deep layer of superficial fascia (Scarpa's) (Membranous)—extends into the perineum as Colles' fascia and the superficial (inferior) layer of the UG diaphragm. The deep layer also extends into the fascia lata.*

B. Deep

1. *Transversalis fascia—covers the deep surface of the transversus muscle*

   i. has extensions (external to the peritoneum) which are named according to their location—e.g., diaphragmatic, iliopsoas, pelvic, femoral

   ii. is attached to the inguinal ligament
   NOTE: Weak transversalis fascia in the floor of Hesselbach's (inguinal) triangle (243) predisposes to a direct inguinal hernia. The boundaries of the triangle are:
   inferior—the inguinal ligament
   superior—the inferior epigastric artery
   medial—the rectus fascia

II. RECTUS SHEATH (233–235)—the aponeurosis of three muscles (external oblique, internal oblique, and transversus). The sheath encloses the rectus muscle, anterior rami of lower 6 thoracic nerves, and the epigastric vessels. The inferior end of the posterior rectus sheath is called the arcuate line.

Rectus abdominis muscle
Anterior layer of rectus sheath
Linea alba
External oblique
Peritoneum
Transversalis fascia
Transversus
Posterior layer of rectus sheath
Internal oblique

A. Above the Arcuate Line

1. *The internal oblique aponeurosis splits, going both in front of and behind the rectus.*

2. *The external oblique aponeurosis goes in front of the rectus.*

3. *The transversus aponeurosis goes behind the rectus.*

B. Below the Arcuate Line—All three aponeuroses go in front of the rectus muscle.

III. MUSCLES (232–234)

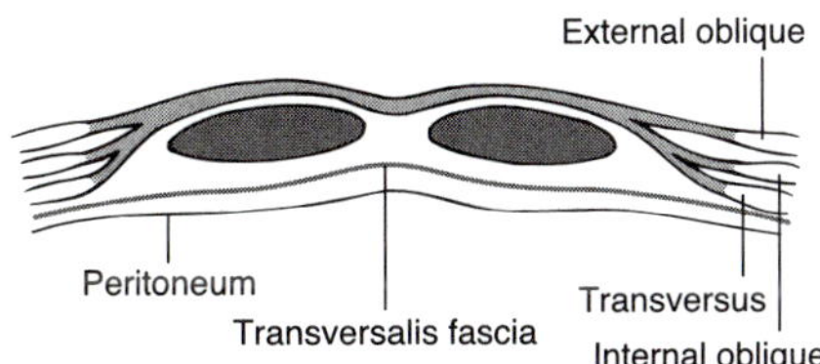

A. External Oblique—innervated by the intercostal and iliohypogastric nerves

1. *Extends from the lower ribs to linea alba, iliac crest, inguinal ligament, and pubis*

2. *Superficial inguinal ring—an opening in the external oblique aponeurosis which transmits the spermatic cord or round ligament*

3. *The inguinal ligament (its lower aponeurotic border)*

   i. also contributes to the lacunar ligament which continues on to the pubic periosteum as the pectineal ligament

   ii. attached to the fascia lata of the thigh

B. Internal Oblique—innervated by intercostal, ilioinguinal, and iliohypogastric nerves

1. *Has lateral origins (lumbar fascia, iliac crest, and inguinal ligament) and medial insertions (ribs, rectus sheath, and pubis)*

2. *The internal oblique aponeurosis joins the transversus to form the* ***conjoint tendon*** *which proceeds to the rectus sheath, pubic crest, and pectineal line.*

3. *Continues on as part of the cremaster muscle*

C. Transversus—innervated by thoracic, ilioinguinal, and iliohypogastric nerves

1. *Extends from costal cartilages of the lower ribs, iliac crest, lumbar fascia, and lateral aspects of the inguinal ligament to the rectus sheath and lacunar ligament*

2. *Aponeurosis joins the internal oblique to form the conjoint tendon*

D. Rectus—innervated by lower intercostals

1. *Extends from lower costal cartilages to the xiphoid and pubis*

2. *The lateral margin of the rectus is called the* ***linea semilunaris (semilunar line).***

IV. ARTERIES (234, 236, 238)

A. Above the Umbilicus—Part of the arterial supply to the anterior abdominal wall is from the superior epigastrics (they enter the rectus sheath) and the lower posterior intercostals.

B. Below the Umbilicus—2 arteries from the external iliac and 3 from the femoral

1. *From the external iliac:*

   i. inferior epigastric—enters the rectus sheath at the arcuate line—raises a peritoneal fold (the lateral umbilical fold)

      a. external spermatic (cremasteric)

      b. suprapubic

ii. deep circumflex iliac—runs parallel to the inguinal ligament. It also supplies the inferior portion of the flanks.

2. *From the femoral:*

i. superficial epigastric—runs toward the umbilicus

ii. superficial circumflex iliac

iii. superficial external pudendal

C. **Anastomoses**—superior and inferior epigastrics

V. VEINS (234, 236, 239)

A. Above the Umbilicus (drainage ultimately into the SVC)

1. *Superior epigastric to the internal thoracic*
2. *Thoracoepigastric to the lateral epigastric to the axillary*

B. Around the Umbilicus—The paraumbilical veins drain into the portal system via the ligamentum teres.

Paraumbilicals have anastomoses with the epigastric veins.

C. Below the Umbilicus—(ultimately into the IVC)

1. *Inferior epigastric into the external iliac*
2. *Superficial circumflex iliac and superficial epigastric into the greater saphenous*

VI. LYMPHATICS (Deep tissues may have additional drainage to lumbar and iliac nodes.)

A. Above the Umbilicus—to the axillary nodes

B. Below the Umbilicus—to epigastrics and superficial inguinal nodes

VII. NERVES (T7-L1) (237, 240, 241)

A. Intercostals (T7–T11) and Subcostals (T12)—course inferiorly and medially down the anterior abdominal wall between the internal oblique and the transversus muscles. The intercostal nerves enter the rectus sheath to supply the rectus muscle.

B. Iliohypogastric (T12-L1)—emerges thru the psoas muscle and goes thru the quadratus lumborum and internal and external oblique muscles to supply the skin above the inguinal ligament and symphysis pubis

C. Ilioinguinal (L1)—follows a slightly lower course than the iliohypogastric, penetrating the internal oblique muscle beneath the aponeurosis of the external oblique to enter the inguinal canal, emerging thru the superficial inguinal ring to supply the skin of the inner thigh and anterior scrotum/labium

VIII. LIGAMENTS (243, 244)

A. Inguinal (Poupart's)—a continuation of the external oblique. It extends between the anterior superior iliac spine and the pubic tubercle.

B. Lacunar (Gimbernat's)—a continuation of the inguinal ligament and the lacunar ligament onto the inferior pubic ramus

C. Pectineal (Iliopectineal) (Cooper's)—covers the pectineal line and extends onto the pubic periosteum. (Some texts refer to the pectineal ligament as a lateral extension of the lacunar ligament and others as a lateral extension of the pectineal portion of the inguinal ligament.)

D. Conjoint Tendon (Inguinal Falx)—a continuation of the aponeurosis of the internal oblique and transversus

The ligaments attached to the bony pelvis are frequently used to provide support in the repair of inguinal and pelvic hernias. Sutures may be incorporated into the inguinal ligament and its extensions (lacunar and pectineal) as part of an inguinal hernia repair.

IX. INGUINAL CANAL (242, 243, 245)—a passageway for the spermatic cord (and its coverings) in the male and the round ligament in the female (sometimes called the "Canal of Nuck" in the female)

A. Embryology (360)

1. *Descent of gubernaculum—a connective tissue layer superficial to the peritoneum. The descent of the testis follows the course of the gubernaculum.*

2. *Processus vaginalis—a peritoneal evagination which precedes the descent of the testes into the scrotum*

   i. acquires coverings which, in the male, remain as coverings for the spermatic cord:

      a. internal spermatic fascia from the transversalis at the deep inguinal ring

      b. cremasteric fascia from internal oblique

      c. external spermatic fascia from external oblique at the superficial inguinal ring

   ii. It has a "remnant" in the scrotum called the tunica vaginalis.

B. Walls

1. *Anterior—aponeurosis of external oblique*

2. *Posterior—transversalis fascia and the medial third of the conjoint tendon (local fusion of aponeurosis of internal oblique and transversus)*

3. *Inferior (floor)—inguinal ligament (Poupart's ligament)*

4. *Superior (roof)—transversus abdominis and internal oblique*

C. The Inguinal Rings (243, 245)

1. *Superficial (triangular in shape)*
    - i. boundaries
        - a. pubic crest
        - b. medial crus of the inguinal ligament
        - c. lateral crus of the aponeurosis of the inguinal ligament
    - ii. transmits the spermatic cord

2. *Deep inguinal ring*
    - i. The deep ring lies halfway between the pubic symphysis and the anterior superior iliac spine and is located lateral to the inferior epigastric artery.
    - ii. an opening in the transversalis fascia
    - iii. The spermatic cord begins at the deep ring.

D. Spermatic Cord—Male (244, 245)

1. *Coverings—spermatic fascia and cremaster fascia. (The cremaster muscle is from the internal oblique.)*

2. *Contents*
    - i. vas deferens (ductus deferens) and its artery
    - ii. testicular vessels—spermatic arteries and veins
        - a. spermatic artery—from the aorta
        - b. cremasteric artery—from the inferior epigastric
        - c. spermatic veins—from the pampiniform plexus
    - iii. remnants of the processus vaginalis
    - iv. genital branch of the genitofemoral nerve—to the cremaster
    - v. autonomic nerves
    - vi. lymphatics

E. Round Ligament—Female (formed by the gubernaculum). The processus vaginalis disappears; there is no female equivalent to the tunica vaginalis. Only the round ligament and the ilioinguinal nerve are in the female inguinal canal. The round ligament extends from the uterus to the labium.

Surface Anatomy of the Anterior Abdominal Wall

I. SURFACE LANDMARKS (231, 232)

A. Costal Margin—formed anteriorly by the costal cartilages of ribs 7–10. (Its lowest margin is the 10th costal cartilage which lies at the level of L3.)

B. Anterior Superior Iliac Spine—at the junction of the iliac crest and the anterior border of the ala of the ilium

C. Posterior Superior Iliac Spine

D. Iliac Crest—that portion of the ilium between the anterior and posterior iliac spines. (Its high point lies at the level of L4.)

E. Tubercle of the Crest (Iliac Tubercle)—lies posterior to the anterior superior iliac spine on the crest

F. Symphysis Pubis—the joint between the pubic bones

G. Pubic Crest—the ridge on the superior surfaces of the pubic bones

H. Pubic Tubercle—a projection on the superior surface of the pubis

I. Inguinal Ligament—a ligamentous continuation of the external oblique. It lies between the anterior superior iliac spine and the pubic tubercle and forms the floor (inferior wall) of the inguinal canal.

J. Superficial Inguinal Ring—an opening in the aponeurosis of the external oblique. It transmits the spermatic cord.

Indirect inguinal hernias begin at the deep inguinal ring and may extend thru the superficial inguinal ring.

K. Linea Alba—formed by the aponeuroses of the external oblique, internal oblique, and transversus. It is located in the midline between the xiphoid and pubic symphysis.

A widened linea alba is called *diastasis recti*. It is usually the result of stretching of the abdominal wall during pregnancy. Epigastric hernias are located in the linea alba. They are usually supraumbilical and may contain a portion of the omentum.

L. Umbilicus—in the linea alba

M. Rectus Muscles—run vertically on either side of the linea alba

N. Linea Semilunaris—the lateral edge of the rectus

II. PLANES (251)

A. Horizontal

1. *Transpyloric—halfway between the jugular notch and the pubic symphysis (L1 level). This plane passes thru the gallbladder, kidneys, neck of the pancreas, and pylorus.*

2. *Subcostal—at the inferior margin of the 10th costal cartilage (L3 level)*

3. *Transumbilical—(L3–L4 level)—thru the level of the umbilicus*

4. *Transtubercular (Intertubercular) plane—between the iliac tubercles (crosses L5 posteriorly)*

5. *Interspinous—between the anterior superior iliac spines*

B. Vertical

1. *Midinguinal (Midclavicular)—from the middle of the clavicle to the middle of the inguinal ligament*

2. *Midsternal—in the middle of the sternum*

3. *Lateral sternal—along the lateral aspects of the sternum*

III. REGIONS ACCORDING TO HORIZONTAL AND VERTICAL PLANES (251)

A. Above the subcostal plane (some texts say above the transpyloric plane)

1. *and between the midclavicular lines—**Epigastric***

2. *to the right and left of the epigastric—**Hypochondriac***

B. Between the subcostal and transtubercular (intertubercular) planes

1. *and between the midclavicular (midinguinal) lines—**Umbilical***

2. *and lateral to the midclavicular lines—**Lumbar (right and left)***

C. Below the transtubercular (intertubercular) plane

1. *and between the midclavicular lines—**Pubic***

2. *and lateral to the midclavicular lines—**Inguinal (Iliac) (right and left)***

IV. REGIONS ACCORDING TO THE UMBILICUS (251)

A. Right Upper Quadrant (RUQ)

B. Left Upper Quadrant (LUQ)

C. Right Lower Quadrant (RLQ)

D. Left Lower Quadrant (LLQ)

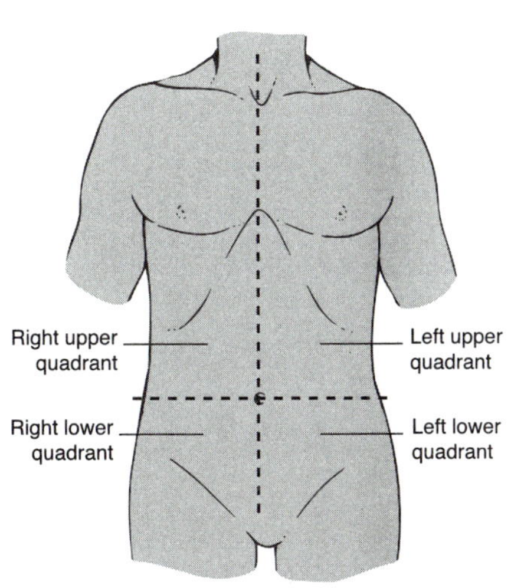

TABLE 5-2. SURGICAL NEUROPATHIES

NERVE	SURGICAL PROBLEM FOLLOWING	POSSIBLE RESULT
Lumbosacral Plexus (T12-S4)	Retractor, suture, abdominal hysterectomy, pelvic sidewall injury—pregnancy	Lumbosacral plexopathy
Ilioinguinal (L1)	Nerve entrapment following a herniorrhaphy or a low and wide lower abdominal incision	Paresthesia of mons pubis, labium/inner scrotum, and medial thigh
Iliohypogastric (T12-L1)	Nerve entrapment following a herniorrhaphy or a low and wide lower abdominal incision	Paresthesia in the hypogastric region
Genitofemoral (L1–L3)	Inguinal herniorrhaphy (injury under the inguinal ligament)	Paresthesia in lateral thigh, labium/scrotum
Genital (L1–L2)	Inguinal herniorrhaphy	Paresthesia of skin of the scrotum and medial thigh
Femoral (L2–L4)	Retractor Dorsolithotomy position	Paresthesia in anterior thigh Weak quads
Lateral Femoral Cutaneous (L2–L3)	Retractor, surgical position, pregnancy	Paresthesia in anterior thigh
Obturator (L2–L4)	Obturator herniorrhaphy, postpartum hematoma, labor	Paresthesia in medial thigh Weakness of thigh muscles
Pudendal (S2–S4)	Pudendal block, forceps, prolonged labor, or pressure from the fetal head during delivery Sacrospinous colpopexy	Perineal neuralgia Levator ani weakness Perineal and gluteal pain
Nerves to anal sphincters	4th-degree laceration	Rectal incontinence
Prostatic plexus	Prostatectomy	Impotence
Nerves accompanying vesical arteries	Radical pelvic surgery	Urinary incontinence
Brachial plexus (C5-T1)	Shoulder dystocia during delivery	Paralysis of upper limb Sensory loss to forearm, hand
Upper trunk of brachial plexus (C5–C6)	Shoulder dystocia during delivery	Paralysis of shoulder muscles and elbow flexors (Erb's palsy)
Brachial plexus (C8-T1)	Upper traction on shoulder during breech delivery	Paralysis of hand (Klumpke's palsy)

Peritoneum

I. LAYERS (252, 254–256)

A. Single

1. *Visceral—covers the viscera*
2. *Parietal—covers the walls of the abdomen and pelvis*

B. Double

1. *Mesentery—connects an organ to the abdominal wall*
   i. dorsal mesentery—suspends the GI tract from the posterior abdominal wall
   ii. ventral mesentery—formed partly by the lesser omentum and falciform ligament. (It is a derivative of the septum transversum.)
2. *Peritoneal ligaments—between two organs* **or** *an organ and the abdominal or pelvic wall*
3. *Omentum—peritoneum leaves curvature of stomach*
   i. lesser omentum—a peritoneal ligament extending from the liver to the lesser curvature of the stomach and the first part of the duodenum. The gastrohepatic ligament forms a major portion of the lesser omentum.
   ii. greater omentum—from the greater curvature of the stomach
      a. gastrophrenic ligament
      b. gastrosplenic ligament
      c. gastrocolic ligament—to the transverse colon

II. RELATIONSHIP OF ORGANS TO PERITONEUM (327–329)

A. Retroperitoneal—The organ is behind the peritoneum.

1. *Abdominal aorta*
2. *Inferior vena cava*
3. *Kidneys*
4. *Pancreas*
5. *Adrenals*
6. *Distal half of the duodenum*
7. *Ascending and descending colon*

B. Intraperitoneal—The organ is mostly covered by the peritoneum.

1. *Liver—except for the bare area*
2. *Stomach*
3. *Proximal half of the duodenum*
4. *Spleen*
5. *Cecum*
6. *Transverse colon*
7. *Body of the uterus*
   i. Peritoneum between the uterus and bladder forms the vesicouterine pouch.
   ii. Peritoneum between the uterus and rectum forms the rectouterine pouch (of Douglas).
8. *Fallopian tubes*

III. PERITONEAL CAVITIES **(255, 256, 329)**

A. Lesser Sac (Omental Bursa)—forms an intraperitoneal space behind the stomach. (It is continuous with the hepatorenal recess also known as the "pouch of Morison.")

1. *Is closed off:*
   i. superiorly—by the diaphragm
   ii. inferiorly—by the transverse colon
   iii. anteriorly—by the stomach, lesser omentum, and gastrocolic ligament

   Because the posterior aspect of the stomach is the anterior aspect of the lesser sac, a posterior perforation of a gastric ulcer may spill gastric contents and concomitant bacterial contamination into the lesser sac.

   The contamination may continue on thru the omental foramen (of Winslow) to the right subhepatic portion of the greater sac and even to other portions of the greater sac.

   iv. posteriorly—by the posterior abdominal wall
2. *Has an opening into the greater sac called the* ***omental (epiploic) foramen of Winslow*** *which has four borders:*
   i. superior—liver
   ii. inferior—1st part of the duodenum

iii. anterior—Its free margin contains the hepatic artery, portal vein, and bile duct.

iv. posterior—inferior vena cava

B. Greater Sac—the **rest** of the peritoneal cavity

The greater sac has recesses, fossae, and pouches. Those in the upper and lower abdomen (i.e., around the liver, bladder, and uterus) tend to collect fluids (serous, purulent, and bloody) whereas those nearer the intestines may entertain and even obstruct a portion of the bowel.

1. *Hepatorenal recess—that portion of the greater sac located between the right kidney, liver, and coronary ligament*
2. *Subphrenic—beneath the diaphragm*
3. *Duodenal recesses—located below the ligament of Treitz*
   i. superior—faces inferiorly
   ii. inferior—faces superiorly
4. *Paraduodenal recess*
5. *Retrocolic recess—located behind the cecum. It may contain the appendix.*
6. *Vesicouterine—between the bladder and uterus*
7. *Rectovesical—between the rectum and bladder*
8. *Rectouterine recess (of Douglas)—between the rectum and uterus—the lowest point of the greater sac in the female*

IV. PERITONEAL LIGAMENTS IN THE ABDOMEN (256, 257)—double layers of peritoneum connecting organs, or an organ to the abdominal or pelvic wall. They may contain blood vessels or their embryological derivatives.

A. Peritoneal Ligaments of the Liver

1. *Coronary ligaments*
   i. anterior—a reflection from the liver to the diaphragm
   ii. posterior—from the right lobe of the liver to the kidney
2. *Triangular ligaments—lateral meetings of the anterior and posterior coronary ligaments*
3. *Falciform ligament—from the umbilicus to the liver (contains the ligamentum teres, a remnant of the umbilical vein covered by peritoneum)*

B. Splenorenal (Lienorenal) Ligament—from the diaphragm to the hilus of the spleen

C. Ligaments of the Lesser Omentum—from the hepatic portal to the stomach

1. *Cholecystoduodenal—the right edge of the lesser omentum*

2. *Gastrohepatic—a major portion of the lesser omentum*

3. *Hepatoduodenal—contains the portal triad*

D. Ligaments of the Greater Omentum

1. *Gastrosplenic (Gastrolienal)—between the stomach and the spleen*

2. *Gastrocolic—from the stomach to the transverse colon*

3. *Gastrophrenic—between the stomach and the diaphragm*

V. PAIN

A. Visceral—Pain fibers are located in the muscular walls of a hollow viscus or in the capsule of a solid viscus and are stimulated by stretch and anoxia. Visceral pain emanates from activation of nerve endings in the walls or capsule of the organ. It is often poorly localized to the midline of the abdomen.

B. Somatic—Somatic pain may be induced by an inflamed organ touching the parietal peritoneum.

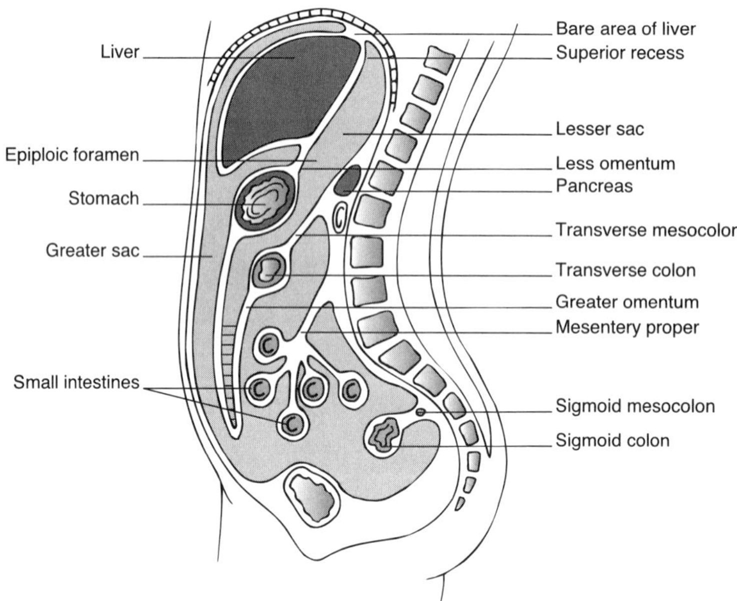

Hernias

I. DEFINITION—A hernia is commonly defined as the protrusion of an organ or other structure thru the wall of the cavity commonly containing it.

A. Congenital Hernias—At birth, the abdominal organ may lie beyond the normal confines of the peritoneal cavity due to incomplete formation of the peritoneal wall.

B. Acquired Hernias—usually occur in developmentally weakened areas of the abdominal wall. The involved area may be one which transmits—i.e., serves as a passageway for—organs, vessels, and nerves. Increased intra-abdominal pressure may then bring about the hernia.

II. INTRA-ABDOMINAL PRESSURE

A. Hernia—A pressure increase within the abdominal cavity may occur suddenly (acute) or over an extended period of time (chronic).

1. *An acute event, such as a sudden severe contraction of the muscles of the abdominal wall as in heavy lifting, may result in an event such as an inguinal hernia.*

2. *A chronic event, such as enlargement of a neoplastic growth, may also cause a pressure increase and result in a hernia.*

B. Abdominal Compartment Syndrome (ACS)—An increase in intra-abdominal pressure may be due to conditions such as gaseous distention of the bowel, an increase in the size of a tumor mass, or even a tight closure of an abdominal incision. Clinically, the abdominal increase in pressure may manifest itself as an increase in central venous pressure or as interference with the contraction of the diaphragm and compromised ventilation.

III. INTRAPERITONEAL "HERNIAS"—entrapment of a viscus in a recess. The recess is the result of an embryologic rotation of the midgut. A mobile organ, usually a portion of the small bowel, may get caught in the recess and become obstructed.

A. Retrocolic hernias—entrapments of intestine behind the mesocolon

B. Paraduodenojejunal hernias—related to the degree of peritoneal rotation and fixation between the DJ flexure and the descending and transverse mesocolon. The *failure of fixation* of the root of the mesentery may later result in small bowel "herniating" thru the opening, resulting in an *intraperitoneal hernia.*

TABLE 5-3. HERNIAS

TYPE AND LOCATION	PREDISPOSING FACTORS	INVOLVED ORGANS
Diaphragmatic (Bochdalek) Lumbocostal (Vertebrocostal) triangle	Incomplete closure of pleuroperitoneal canal	Abdominal viscera into thorax
Parasternal (Morgagni) (Congenital)	Incomplete closure of sternocostal triangle	Omentum or colon
Esophageal (Hiatal Hernia)	Weakness of diaphragm sphincter Short esophagus Increased vagal activity	Abdominal portion of the esophagus and upper stomach into the thorax
Paraesophageal (Hiatal)	Weakness of esophageal sphincter Decreased lower esophageal sphincter tone	Stomach migrates into posterior mediastinum (alongside the esophagus)
Epigastric	Surgery involving linea alba Usually supraumbilical	Omentum
Umbilical—Congenital	Incomplete closure of umbilical ring	Abdominal viscera
Umbilical—Acquired	Pregnancy, tumor, ascites	Abdominal viscera
Semilunar (Spigelian)	Weakness at junction of arcuate line and linea semilunaris	Fat, omentum—usually remain behind muscles
Omphalocele (Congenital)	Failure of bowel to return to abdominal cavity	Gut covered by amnion
Gastroschisis (Congenital)	Defect lateral to umbilicus	Gut not covered by amnion
Incisional	Surgery	Abdominal viscera
Paraduodenal (Congenital)	Malrotation of gut	Small bowel
Indirect Inguinal	Weakness of internal inguinal ring	Small bowel
Inguinal (Congenital)	Failure of separation of peritoneum and processus vaginalis	Gut
Direct Inguinal (passes thru inguinal triangle)	Weakness of muscles medial to inferior epigastric vessels	Small bowel
Obturator	Weakness of obturator foramen (more common in females)	Bowel Obturator nerve
Sciatic (Sacrosciatic) Gluteal	Atrophy of the piriformis muscle	Bowel, bladder, ureter, ovary
Femoral	Weakness of femoral ring (more common in females)	Small bowel into femoral canal
Retrocolic (Internal) (Congenital)	Lack of fusion of abdominal wall and mesentery	Gut

Celiac Artery

I. LEFT GASTRIC (282, 286, 288)—runs in the lesser omentum

A. Esophageal—to the diaphragmatic part of the esophagus

B. Descends from left to right, behind the lesser sac, to supply the lesser curvature of the stomach—**anastomoses** with right gastric

II. SPLENIC (283)—passes along the upper border of the pancreas

A. Short (Posterior) Gastrics—to the superior portion of the greater curvature of the stomach

B. Greater Pancreatic—supplies body and tail of the pancreas

C. Left Gastro-omental—passes between the anterior layers of the omentum and has **anastomoses** with the right gastro-omental

1. *Gastric branches—supply the greater curvature*
2. *Omental branches—to the greater omentum*

D. Splenic branches—extend along the upper border of the pancreas to enter the splenorenal (lienorenal) ligament and end in 5–8 terminal branches to the spleen.

The mobility and tortuosity of the splenic artery make it susceptible to thrombosis, aneurysm, and rupture.

III. COMMON HEPATIC (282, 284, 288)—turns upward in the lesser omentum (anterior margin of the foramen of Winslow) in front of the portal vein

A. Gastroduodenal—arises above the upper border of the 1st part of the duodenum

The gastroduodenal artery may be eroded by the contents of a perforated duodenal ulcer.

1. *Superior pancreaticoduodenal—descends behind the first part of the duodenum to supply the first and second parts of the duodenum and the head of the pancreas*
   ***NOTE: Anastomoses** with inferior pancreaticoduodenal of the superior mesenteric*
2. *Right gastro-omental (Gastroepiploic)—gastric and omental branches—**anastomoses** with the left gastro-omental*
3. *Right gastric—supplies lesser curvature of stomach—**anastomoses** with the left gastric along the lesser curvature*

B. Proper Hepatic

1. *Right hepatic—to the right lobe of the liver*
   i. Cystic—to the gallbladder
2. *Left hepatic—to the left lobe of the liver*

Liver

I. DIVISIONS (272)

A. Lobes—separated by the falciform ligament

1. *Right—has quadrate and caudate portions. An inferior extension of the right lobe is called Reidel's lobe.*

2. *Left—smaller than right*

B. Lobes—divided into **lobules** (272, 273, 275) which contain:

1. *Sinusoids—lined by endothelial cells which are separated from hepatocytes by the space of Disse. Sinusoids contain arterial blood (30%) and portal venous blood (70%).*

2. *Acini—cylinders of liver cells (hepatocytes) around the portal tract with central veins at the periphery*

3. *Portal tracts—branches of the hepatic artery, portal vein, bile ducts, and lymphatics*

4. *Central vein—in each lobule—the first tributary of the hepatic vein—drains into sublobular veins*

5. *Canals—between lobules*

C. Segments (272)—Each has a tributary of the portal vein, a branch of the hepatic artery, and a branch of the bile duct.

II. LIGAMENTS (270)—double layers of peritoneum

A. Peritoneal

1. *Coronary—three sided*

2. *Triangular—left and right, formed by the right and left ends of the coronary ligaments*

B. Falciform—from the umbilicus. It contains the ligamentum teres (a remnant of the umbilical vein).

C. The Gastrohepatic Ligament (lesser omentum) extends to the porta hepatis (hepatic portal) and surrounds the hepatic artery, the portal vein, and the bile ducts.

III. PORTA HEPATIS (270)—A fissure in the liver which contains the portal vein and the hepatic artery, nerves, lymphatics, and ducts. It contains structures which define 2 major triangles:

A. Hepatocystic Triangle—located within the margin of the liver, the common hepatic duct, and the cystic duct

B. Calot's Triangle (284)—located within the cystic artery, the cystic duct, and the common hepatic duct. **(The medial portion of the triangle, near the common hepatic duct, is considered a surgical danger zone.)**

IV. DUCTS (276)—The right and left hepatic bile ducts form the common hepatic duct (in the edge of the lesser omentum).

V. LIVER SURFACES (269, 270)

A. Anterosuperior—convex—related to the right lung

B. Right Lateral—related to the diaphragm

C. Posterior—related to the diaphragm and IVC

D. Inferior—may have impressions for the right kidney and the hepatic flexure

VI. RELATIONSHIPS (269)—The upper border of the liver goes up to the 4th intercostal space anteriorly. The lower border extends to the costal margin.

A. To Right Lobe—gallbladder, duodenum, right colic flexure, right kidney, and suprarenal

B. To Left Lobe—esophagus and stomach

VII. ARTERIES (282)

A. The Common Hepatic Artery—runs in the gastrohepatic ligament. In the porta hepatis, it divides into the right and left hepatics. The right hepatic artery lies posterior to the common hepatic duct. (About 10% of the time it is **anterior**.)

B. Accessory Arteries—include the paraumbilical and occasional small branches from the cystic or mesenteric arteries

VIII. VEINS (274, 290)

A. Portal—enters the liver in the porta hepatis

B. Hepatic Veins—right and left enter the inferior vena cava

IX. LYMPHATICS (298)

A. Hepatic Nodes to celiac nodes

B. Phrenic Lymphatics to parasternal nodes

X. NERVES (309)

A. Sympathetics—increase glycogenolysis and decrease bile production

B. Parasympathetics—increase glycogenolysis and increase bile production

C. Visceral Afferents—Dorsal root ganglia of spinal nerves T5–T9 contain cell bodies of sensory nerves. Other afferents go via the right phrenic nerve to C3, C4, and C5.

Gallbladder

The gallbladder is a hollow viscus which lies against the liver between the right lobe and the quadrate lobe. It concentrates and serves as a reservoir for bile. It is located in the right upper quadrant opposite the tip of the right 9th costal cartilage (where the linea semilunaris crosses the costal margin). The gallbladder contracts under the influence of cholecystokinin (CCK). The mucosa consists of columnar epithelium in folds. The gallbladder is about 3–4 inches long and holds a jigger of bile. It is derived from the hepatic diverticulum of the foregut.

I. DIVISIONS (276)

A. Fundus—If it projects beyond the liver, it is covered on all sides by peritoneum.

B. Body—covered by peritoneum by the sides and inferiorly

C. Infundibulum—**Hartmann's pouch—where stones may get caught**—connected to the duodenum by the right edge of the lesser omentum

D. Neck—Mucosa folds form the "spiral valve of Heister." The neck funnels into the cystic duct.

II. RELATIONSHIPS (276)

A. Superior—the liver

B. Anterior—the abdominal wall

C. Posterior—colon and duodenum

III. CYSTIC ARTERY (276, 288)—from the right hepatic—has superficial and deep branches. There are many variations in the origin of the cystic artery (e.g., common hepatic, superior mesenteric, right gastro-omental, and celiac trunk), its course, and its branching.

IV. CYSTIC VEIN (294)—empties into the portal vein

V. LYMPHATICS (295, 298)—to cystic and hepatic nodes and thence to celiac nodes

VI. NERVES (309)

A. Parasympathetics via the Vagus

B. Sensory fibers (visceral afferents) via the phrenic nerve and greater splanchnics (T5–T9)

An inflamed gallbladder may also irritate the anterior abdominal wall and produce point tenderness.

It may also produce painful sensations at the tip of the scapula.

VII. FORAMEN OF WINSLOW (256)—found by placing a finger along the free margin of the lesser omentum. The finger palpates the common duct, hepatic artery, and portal vein. Superior to the finger is the gallbladder; inferior is the duodenum; posterior is the IVC.

Bile Ducts

I. **HEPATIC DUCTS, RIGHT AND LEFT (276)**—drain the right and left portions of the liver. The right and left hepatic ducts form the common hepatic duct in the porta hepatis.

II. **COMMON HEPATIC DUCT (276)**—The common hepatic duct lies lateral to the hepatic artery in the lesser omentum. The portal vein lies posterior to the duct and artery.

III. **CYSTIC DUCT (276)**—contains the "spiral valve of Heister," a spiral fold (i.e., an infolding of the mucous membrane duct wall). In some texts the portion of the cystic duct containing the "spiral valve" is listed as the neck of the gallbladder. The cystic duct lies parallel to the hepatic duct. It passes in the free edge of the lesser omentum with the hepatic artery and portal vein. The length and course of the cystic duct may vary. This presents a risk of surgical injury to the junction of the common hepatic and cystic ducts. The cystic duct joins the common hepatic duct to form the common bile duct.

IV. **COMMON BILE DUCT (278, 279)**—formed by the union of the cystic and common hepatic ducts. The common bile duct descends behind the duodenum and anterior to the inferior vena cava. It passes thru the posterior pancreas to join the pancreatic duct and enter the second part of the duodenum at the ampulla of Vater. Occasionally the common bile and pancreatic ducts open separately into the duodenum. An impacted gallstone at the ampulla of Vater may also block the pancreatic duct and result in chronic pancreatitis. This may be obviated if a separate pancreatic duct is present.

V. **DUCTAL ARTERIES (276, 282)**—form anastomoses on the ductal walls

A. Hepatic ducts—The right hepatic artery travels on the inferolateral side of the right hepatic duct.

B. Common Hepatic Duct—The pancreaticoduodenal artery has branches to the lateral sides of the duct.

C. Cystic Duct and Gallbladder—The cystic artery sends 2 main branches along the margins of the duct and gallbladder. The cystic artery arises, from the right hepatic, in an angle between the cystic duct and the common hepatic duct. When the cystic artery arises to the left of the common hepatic duct, it usually crosses it anteriorly. (The right hepatic artery may lie close to the cystic duct.)

D. Common Bile Duct—supplied by the pancreaticoduodenal or occasionally the gastroduodenal artery. The superior mesenteric or celiac trunk sends a branch to the retroduodenal portion of the common bile duct.

Stomach

I. DIVISIONS (258, 259)

A. Fundus—above the level of the cardiac (esophageal) opening

1. *Contains the gastric pacemaker*
2. *Columnar epithelium—Glands open into bases of gastric pits.*

B. Body—located between the fundus and antrum

1. *Columnar epithelium—Glands open into bases of gastric pits.*
2. *Specialized secretory cells*
   i. parietal cells—in upper ½ of glands—produce HCl (stimulated by the Vagus) and intrinsic factor—contain histamine receptors which stimulate acid production
   ii. chief cells—in lower ½ of glands—produce pepsin
3. *Rugae (folds)—of mucosa*

C. Antrum—the distal ⅓ of the stomach. It is smooth (no rugae).

1. *Glands—antral or pyloric glands—secrete mucus*
2. *"G" cells—produce gastrin. Gastrin release is stimulated by the Vagus secondary to antral distention or amino acids in the lumen.*
3. *"D" cells—produce somatostatin*
   NOTE: *The antrum is sometimes divided into two parts called the pyloric antrum and the pyloric canal.*

II. OPENINGS (259)

A. Cardiac Orifice (Cardia)—the esophageal opening into the stomach. Muscles at the esophagogastric junction form the cardiac (esophageal) sphincter.

B. Pyloric Orifice—opening into the duodenum via the pyloric sphincter

III. CURVATURES (258)

A. Greater Curvature

1. *Cardiac notch (Incisura)—an acute angle between the esophagus and the fundal portion of the greater curvature*
2. *Gives rise to the greater omentum*
3. *Gives rise to the gastrosplenic (gastrolienal) ligament (the part of the greater omentum which extends to the spleen)*
4. *Receives blood supply from gastro-omental arteries*

B. Lesser Curvature

1. *Has an angle called the* **angular incisura** *(incisura angularis) located between the body and the antrum*
2. *Gives rise to the lesser omentum which extends to the liver*
3. *Receives blood from the gastric arteries*

   A tear in the mucosa and submucosa of the lesser curvature due to severe vomiting is called Mallory-Weiss syndrome.

IV. WALLS (260)

A. Longitudinal Muscle—superficial layer—more muscle fibers along the curvatures

B. Circular Muscle Layer—contributes to the pyloric sphincter. At the pyloric sphincter, it is thick.

Excess thickening of the muscle may be responsible for congenital pyloric stenosis and a narrowing of the pyloric lumen, resulting in progressively increasing episodes of vomiting. If obstruction results, surgery is frequently indicated.

C. Oblique Muscle Layer—contributes to the pyloric sphincter

D. The myenteric (Auerbach's) plexus is located between the longitudinal and circular muscle layers.

V. LIGAMENTS (255, 257)

A. Gastrosplenic—part of the greater omentum

B. Gastrohepatic—part of the lesser omentum

C. Gastrocolic—part of the greater omentum

VI. RELATIONSHIPS (256)

A. Anterior—lung, diaphragm, and liver

B. Posterior—kidney and suprarenal, spleen, pancreas, and colon

VII. ARTERIES (282, 283, 288)—from the celiac artery

A. Along the Greater Curvature

1. *Left gastro-omental (gastroepiploic)—from the splenic*
2. *Right gastro-omental (gastroepiploic)—from the gastroduodenal of the hepatic*
   i. pyloric branch
3. *Short gastrics—from the splenic—supply the superior (fundal) portion of the greater curvature*

B. Along the Lesser Curvature—have **anastomoses**

1. *Left gastric*

2. *Right gastric—from the hepatic*

➢VIII. VEINS (290)

A. Gastrics—drain into the portal vein. The left gastrics **anastomose** with the esophageal.

B. Left Gastro-omental (Gastroepiploic)—into the splenic

C. Right Gastro-omental (Gastroepiploic)—into the superior mesenteric

D. Short Gastrics—into the splenic

IX. OMENTAL LYMPHATICS (299)—follow arteries to celiac nodes

A. Nodes in the Greater Omentum (Gastro-omental)—located along the greater curvature

1. *Subpyloric and inferior gastric nodes—to gastroduodenal (pyloric) or celiac nodes*

2. *Pancreaticosplenic nodes—drain the fundus of the stomach*

B. Nodes in the Lesser Omentum

1. *Right gastric (Superior gastric) (Supragastric)—to the pyloric (gastroduodenal) nodes*

2. *Suprapyloric—to the pyloric nodes*

➢ X. LYMPHATICS (295)

A. Fundus—pancreaticosplenic

B. Pylorus—pyloric nodes

1. *Suprapyloric—in the lesser omentum*

2. *Subpyloric—in the greater omentum*

C. Lesser curvature—to left and right gastric nodes

D. Greater curvature—to gastro-omental nodes located in the greater omentum

➢ XI. NERVES (301–303)

A. Sympathetic—via the celiac ganglion—inhibit production of gastric juices, decrease motility, and contract the pyloric sphincter—preganglionics, via splanchnic (mostly thoracic) synapse in the celiac plexus

B. Parasympathetic (Vagus) nerves—stimulate production of gastric juices—synapse in the stomach wall

1. *Anterior trunk*

   i. anterior gastric division—supplies the anterior wall of the stomach

   ii. hepatic division—supplies the proximal duodenum

2. *Posterior trunk—supplies the posterior gastric wall*

C. Visceral Afferents—via the celiac plexus (pain fibers)—travel along with sympathetics. Stomach pain is referred to the epigastric area (T5–T6).

D. ENS—regulates gastric emptying via VIP and CCK

XII. STOMACH ABNORMALITIES

A. **Acute gastric volvulus—characterized by severe upper abdominal pain radiating into the neck and back. It may occur in a paraesophageal hernia.**

B. **Zollinger-Ellison Syndrome—a neuroendocrine (non-B islet cell tumor) found in the duodenum, pancreas, or jejunum. The tumor secretes excess gastrin which in turn is responsible for excess parietal acid secretion and frequent peptic ulcer perforation. The gastrinoma stimulates peristalsis and results in diarrhea.**

C. **Gastric Ulcer Perforation—Stomach contents in an anterior perforation may go into the greater sac while a posterior perforation goes into the lesser sac.**

Spleen

The spleen (281) is an encapsulated lymphatic immunologic filter and factory located under the diaphragm near the 10th rib. Although its relationships vary, it may be related anteriorly to the stomach, the tail of the pancreas, and left colic flexure. The spleen has a tortuous artery and is not strongly supported. As a result, it is quite movable and prone to injury (rupture). The size of the spleen varies (average about 5 by 3 inches), and it weighs between 80 and 300 grams.

I. COMPOSITION

A. Capsule—dense connective tissue sending out trabeculae which divide the pulp into incomplete compartments

B. White Pulp—nodular accumulations of T and B lymphocytes, macrophages, and special "antigen-presenting" cells surrounding arterioles (Malpighian corpuscles)

C. Red Pulp (Splenic Cords)—venules and phagocytes in a reticular network

➢ II. LIGAMENTS (281)—**form the splenic pedicle (4 layers) which connects the spleen to the kidney and stomach**

A. Lienorenal—contains splenic vessels and the tail of the pancreas

B. Gastrosplenic—to the greater omentum

1. *Contains short gastric vessels*
2. *Contains the left gastro-omental vessels*
3. *Divides at the splenic hilus*

➢ III. SPLENIC ARTERY (281, 283, 288)—**a branch of the celiac. It enters the spleen as 5–8 terminal branches.**

The tortuosity of the splenic artery may lead to an aneurysm, thrombosis, or even (rarely) rupture.

➢ IV. SPLENIC VEIN (290)—**runs behind the pancreas. It is formed at the hilus and is joined by the left gastro-omental (gastroepiploic) and passes behind the pancreas and below the splenic artery to join the superior mesenteric and become the portal vein.**

➢ V. LYMPHATICS (299)—**Lymphatic nodules in the hilus empty into the pancreaticosplenic and celiac nodes.**

➢ VI. NERVES (301)

A. Sympathetic—contract the spleen, causing increase of blood into the circulation.

B. Parasympathetic—Vagus

Superior Mesenteric Artery

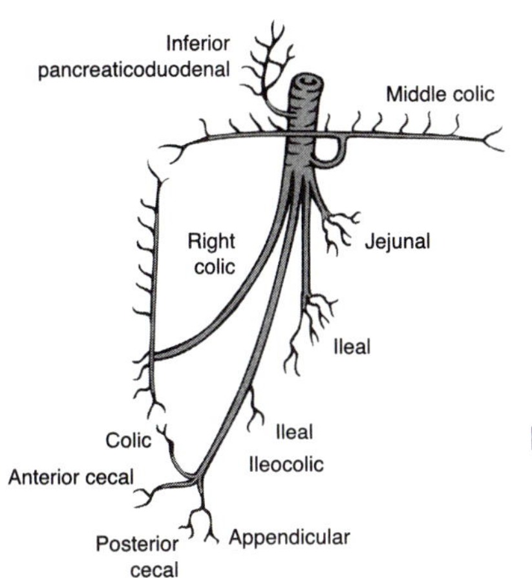

The superior mesenteric artery arises from the aorta, just below the celiac, proceeds behind the pancreas, then forward between the uncinate process of the pancreas and the 3rd part of the duodenum (the bend makes it susceptible to infarction) and thence to enter the mesentery. It supplies the 3rd and 4th parts of the duodenum, the jejunum, the ileum, and the colon up to the distal portion of the transverse colon.

The main branches of the superior mesenteric artery (283–289) are:

I. INFERIOR PANCREATICODUODENAL (285)

A. Supplies the head of the pancreas and a portion of the duodenum

B. Has anterior and posterior branches to the duodenum and the head of the pancreas. They **anastomose** with the superior branch of the gastroduodenal.

➢ II. MIDDLE COLIC (287)**—begins just below the pancreas to supply the transverse colon including hepatic and splenic flexures. It runs in the transverse mesocolon.**

A. Right branch—to the right transverse colon

B. Left branch—to the left transverse colon

➢ III. RIGHT COLIC (286, 287)**—arises retroperitoneally along with the ileocolic artery—supplies the ascending colon**

A. Ascending Branch—**anastomoses** with the right branch of the middle colic near the hepatic flexure

B. Descending Branch—**anastomoses** with the colic of the ileocolic

➢ IV. ILEOCOLIC (286, 287)**—supplies the terminal ileum, cecum, and appendix**

A. Anterior Cecal—supplies the anterior surface of the cecum

B. Posterior Cecal—supplies the posterior surface of the cecum

1. Appendicular (Appendiceal)—to the appendix

C. Ileal—to the terminal ileum—**anastomoses** with the superior mesenteric

D. Colic—**anastomoses** with the descending branch of the right colic

➢ V. JEJUNAL AND ILEAL ARTERIES (286)

A. Each divides into 2 branches

B. Terminal branches go to the intestinal muscularis

Portal System

I. **PORTAL VEIN (292–294)—begins behind the neck of the pancreas by the union of the superior mesenteric and splenic veins. It passes behind the first part of the duodenum and enters the lesser omentum where it receives the gastric and pancreaticoduodenal veins. In the free border of the lesser omentum, it lies posterior to the bile duct and hepatic artery. In the porta hepatis, it divides into right and left branches. The portal vein ultimately receives blood from the pancreas, the spleen, and the alimentary tract from the lower esophagus to the upper end of the anal canal (the organs supplied by the celiac and mesenteric arteries).**

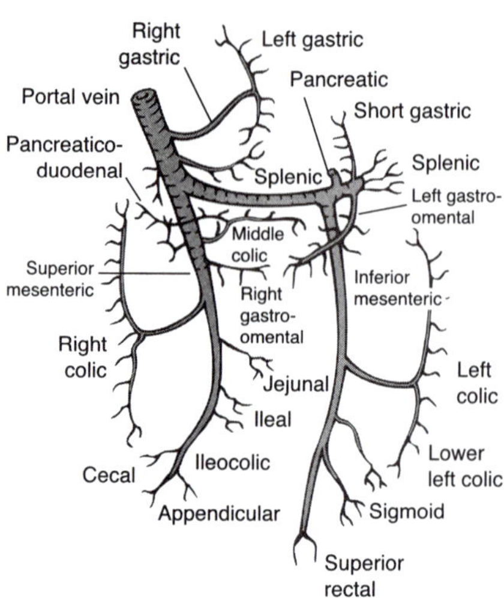

A. Splenic (290)

1. *Inferior mesenteric—a continuation of the superior rectal. It drains the descending colon, sigmoid colon, and proximal rectum.*
2. *Short gastrics—drain the superior portion of the greater curvature of the stomach*
3. *Left gastro-omental—drains the greater curvature*

B. Superior Mesenteric (291)—receives blood from the second part of the duodenum to the right half of the transverse colon. Its course parallels branches of the superior mesenteric artery plus the right gastro-omental and pancreaticoduodenal arteries.

C. Cystic (294)—may go directly into the liver

D. Pancreaticoduodenals (290, 293, 294)—from the pancreas and duodenum

II. **PORTAL-SYSTEMIC ANASTOMOSES (293)—Areas which are drained by tributaries of both the portal vein and the vena cava have portal-systemic anastomoses. Because the portal system does not have valves, an increase in portal-venous pressure is transmitted throughout all of the portal vein tributaries in a retrograde fashion.**

Pressure differences in the areas of portal-systemic venous anastomoses produce venous dilation. These veins may enlarge and can even rupture with a large loss of blood.

TABLE 5-4.

PORTAL TRIBUTARIES	SYSTEMIC VEINS	CLINICAL CONDITION
Esophageal veins (of left gastric)	Esophageal (of azygos)	Esophageal varices
Superior rectal veins	Middle rectal	Hemorrhoids
Veins in the falciform ligament	Superficial epigastrics	"Caput Medusae"
Celiac and Splenic veins	Renal	
Veins of the bare area of the liver	Inferior phrenic	
Splenic and Pancreatic veins	Left renal	
Splenic and Colic veins	Lumbar	

Duodenum

I. GENERAL—The duodenum is about 10 inches long. It extends from the pylorus to the duodenojejunal flexure at the ligament of Treitz. The duodenum is C shaped around the head, neck, and body of the pancreas.

➢ A. Arteries (283–285)

1. *Superior pancreaticoduodenal—from the gastroduodenal—has anterior and posterior branches—supplies proximal part of duodenum—**anastomoses** with the inferior pancreaticoduodenal*
2. *Inferior pancreaticoduodenal—from the superior mesenteric—supplies the distal part of the duodenum*
3. *Other branches of the gastroduodenal including the supraduodenal and retroduodenal may supply the superior portion of the duodenum.*

➢ B. Veins (290)—portal tributaries (including the superior mesenteric and those veins accompanying the arteries)

➢ C. Lymphatics (295, 296)—gastroduodenal (pyloric) and superior mesenteric

➢ D. Nerves (301–303)

1. *Sympathetic—decreases motility*
2. *Parasympathetic (Vagus)—increases motility*
3. *Visceral afferents—greater splanchnics via sympathetic trunk to T5–T9*

➢ **II. SUPERIOR—1ST PART (261)—length 2 inches—vertebral level L1**

A. Arteries

1. *Superior pancreaticoduodenal from gastroduodenal from hepatic*

B. Veins—superior pancreaticoduodenal

C. Lymphatics—pancreaticoduodenal to pyloric (gastroduodenal) to celiac nodes

D. Nerves—celiac plexus

➢ **III. DESCENDING—2ND PART (261)—length 4 inches—vertebral level L2**

A. Arteries—superior pancreaticoduodenal from the gastroduodenal

B. Veins—superior pancreaticoduodenal and superior mesenteric

C. Lymphatics—gastroduodenal (pyloric) and pancreaticoduodenal

D. Nerves—superior mesenteric plexus

➢ IV. HORIZONTAL (INFERIOR)—3RD PART (261)—length 4 inches—vertebral level L3

A. Arteries—inferior pancreaticoduodenal from superior mesenteric

B. Veins—superior mesenteric and pancreaticoduodenal

C. Lymphatics—pancreaticoduodenal and superior mesenteric

D. Nerves—superior mesenteric plexus

➢ V. ASCENDING—4TH PART (261)—(terminates at the duodenojejunal flexure, where a fixed part and a freely movable part of the intestine are joined)—length 1 inch—vertebral level L2–L3

A. Arteries—inferior pancreaticoduodenal from superior mesenteric

B. Veins—superior mesenteric and pancreaticoduodenal

C. Lymphatics—superior mesenteric and pancreaticoduodenal

D. Nerves—superior mesenteric plexus

➢ VI. FOSSAE (261)—recesses formed by developmental intestinal-mesenteric rotation

A. Paraduodenal—lies to the left of the duodenojejunal flexure. It opens to the right.

B. Duodenojejunal—formed by peritoneal folds

1. *Superior (recess)—lies to the left of the duodenojejunal flexure and faces inferiorly*

2. *Inferior (recess)—faces superiorly—lies in front of L3 vertebra*

C. Retroduodenal—between the superior and inferior recesses

VII. HISTOLOGY

A. Mucosa—tall villi with crypts of Lieberkühn

Duodenal endocrine cells of the ENS release cholecystokinin (CCK) which contracts the gallbladder and relaxes the Sphincter of Oddi

Somatostatinomas—duodenal or pancreatic tumors

1. *inhibit islet cell insulin → diabetes*

2. *inhibit gallbladder emptying → gallstones*

3. *inhibit pancreatic enzymes → steatorrhea*

B. Muscularis Mucosa

C. Submucosa—Brunner's glands

D. Muscle—circular and longitudinal

TABLE 5-5. DUODENUM

	SUPERIOR—1ST	DESCENDING—2ND	HORIZONTAL—3RD	ASCENDING—4TH
RELATIONSHIP				
Anterior	Liver Gallbladder	Liver Gallbladder Transverse Colon	Small Intestine Superior Mesenteric Artery	Jejunum Root of mesentery
Posterior	Bile Duct Portal Vein IVC Gastroduodenal Artery	Right Kidney Ureter Psoas major IVC	Psoas Ureter Aorta IVC	Left Psoas Aorta Left renal artery Left spermatic/ovarian vessels
Superior	Lesser Sac Right Gastric Artery		Pancreas	Body of the pancreas Jejunum
Inferior	Pancreas			
Lateral		Right lobe liver Ascending colon		Vertebral column Jejunum
Medial		Head of Pancreas		
PART				
ARTERIES	Superior Pancreaticoduodenal	Superior pancreaticoduodenal	Inferior Pancreaticoduodenal	Inferior Pancreaticoduodenal
NERVES	Celiac Plexus	Celiac Plexus	Superior Mesenteric Plexus	Superior Mesenteric Plexus
LIGAMENTS	Hepatoduodenal (part of the lesser omentum)			Suspensory Ligament (of Treitz)
LENGTH	1–2 inches	3–4 inches	4 inches	1 inch
VERTEBRAL LEVEL	L1	L2	L3	L2–L3
OPENINGS	Pylorus	Ampulla of Vater	2nd and 4th parts	Jejunum
PERITONEUM	Front and back	Front	Front	Front
MUCOSA	Smooth	Circular folds		

Pancreas

The pancreas is a retroperitoneal gland derived from the foregut and midgut.

I. RELATIONSHIPS OF PANCREATIC DIVISIONS (279)

A. Head—in the concavity of the duodenum and attached to the transverse mesocolon (lies on the right due to gut rotation)

1. *The uncinate process, from the lower part of the head, hooks behind the superior mesenteric vessels.*
2. *The superior mesenteric and portal veins lie between the head and neck because the pancreas is embryologically derived from two "buds" which fuse.*

B. Neck—in front of the portal vein and behind the pylorus

C. Body—has 3 surfaces

1. *The splenic artery extends along the upper border of the body and tail*
2. *Behind the body lies the aorta and left kidney.*

D. Tail—lies in the splenorenal (lienorenal) ligament and may contact the spleen

II. FREQUENT RELATIONSHIPS OF THE PANCREAS TO OTHER STRUCTURES (279)

A. Anterior—colon and stomach (pylorus)

B. Posterior—bile duct, portal vein, splenic vein, inferior vena cava, renal vein, crura of diaphragm, aorta, superior mesenteric artery, left adrenal gland, and kidney

C. The Common Bile Duct—lies on the anterior surface of the pancreas

D. That Portion of the Pancreas Below the Transverse Colon—covered with peritoneum. This portion is related to the duodenum, jejunum, and left colic flexure.

III. GLANDS (310)

A. Exocrine—hydrolytic enzymes enter pancreatic duct

1. *Acini—composed of epithelial cells which converge to form a lumen. They produce amylase, lipase, trypsinogen, and bicarbonate.*
2. *Lobule—composed of several acini*

B. Endocrine

1. *Islets of Langerhans*

   i. Alpha cells—produce glucagon

   ii. Beta cells—produce insulin

   iii. Delta cells—produce somatostatin

2. *Vasoactive intestinal polypeptide (VIP)*

   Vipomas are rare ENS-related endocrine tumors causing excess bowel secretions and bowel relaxation resulting in diarrhea, achlorhydria, and hypokalemia.

3. *Pancreatic polypeptide (PP)*

IV. DUCTS (279, 280)

A. Chief Pancreatic (Duct of Wirsung)—extends laterally (to the right) to join the bile duct and empty into the duodenum at the ampulla of Vater

B. Accessory Pancreatic (Duct of Santorini)—drains the superior part of the pancreatic head—empties into the Chief pancreatic duct or may open separately into the duodenum

V. ARTERIES (283–285)

A. Pancreaticoduodenals—supply the head

1. *Superior pancreaticoduodenal (from the gastroduodenal)—passes between the head of the pancreas and the duodenum*

2. *Inferior pancreaticoduodenal (from the superior mesenteric)—has **anastomoses** with 1.*

B. Splenic—supplies the body and tail—the major supply

VI. SPLENIC VEIN (290)—joins the superior mesenteric (behind the neck) to form the portal vein.

VII. LYMPHATICS (299)—pancreaticosplenic, pyloric, and superior mesenteric nodes

VIII. NERVES (310)

A. Sympathetics—decrease enzyme production

B. Parasympathetics—increase endocrine secretions

C. Visceral afferents—Pain fibers run thru thoracic splanchnic nerves (greater, lesser, and least) and thru the celiac plexus. They enter spinal cord segments T6–T10. Therefore, pain from the pancreas may be felt in the upper abdomen.

Jejunum and Ileum

The jejunum begins at the ligament of Treitz and then gradually blends into the ileum. The combined length of the jejunum and ileum is about 22 feet. They are suspended by a mesentery. A diverticulum (Meckel's) occasionally occurs within the terminal 2 feet of the ileum. This diverticulum is the result of persistence of a portion of the omphalointestinal (vitello) duct. Meckel's may be lined with gastric mucosa. This diverticulum is a common cause of severe bleeding of the lower intestinal tract.

➢ **I. SUPERIOR MESENTERIC ARTERY (286)—arises from the aorta (behind the pancreas) and then passes in front of the pancreas and left renal vein**

A. Jejunum

1. *Jejunal branches form arcades (arches) which form one tier of vasa recta (straight arteries or arteriae rectae).*
These arteries are longer than those of the ileum.

B. Ileum—supplied by the ileal branch of the superior mesenteric

1. *Last ileal branches* ***anastomose*** *with the ileocolic artery which may supply the terminal ileum*

2. *Greater number of arcades (arches) than the jejunum but they are shorter. Terminal tiers of the arteries form vasa recta (straight arteries or arteriae rectae).*

➢ **II. VEINS (291)—Superior mesenteric (which unites with the splenic to form the portal vein)**

➢ **III. LYMPHATICS (296)—to superior mesenteric nodes**

➢ **IV. NERVES (304)—via splanchnics and the celiac plexus**

A. Sympathetic

1. *Superior mesenteric plexus*

2. *Synapses in superior mesenteric ganglia*

B. Parasympathetic—via superior mesenteric plexus from the posterior vagal trunk—increases motility

C. Visceral Afferents—via splanchnics T8–T12—therefore, pain may be referred to the umbilical area

D. ENS—The Migratory Motor Complex (MMC) controls the peristaltic reflex which regulates circular smooth muscle contraction and relaxation. This action moves food particles and prevents bacterial overgrowth.

TABLE 5-6. JEJUNUM AND ILEUM COMPARISONS

PART	JEJUNUM	ILEUM
Portion of Small Intestine	Upper ⅖	Lower ⅗
General location	LUQ	RLQ and Hypogastrium
Wall thickness	Greater	Lesser
Caliber	Greater	Lesser
Mucosa	Thicker	Thinner
Villi	Greater number Finger-like Surface contains columnar absorptive epithelial cells and Goblet cells Crypts of Lieberkühn open between bases of villi Crypts contain Paneth cells	Lesser number Taller and more slender Greater number of Goblet cells
Plicae	Larger	Smaller
Mesentery length	Shorter	Longer
Mesentery attachment	Above and to left of the aorta	Below and to right of the aorta
Mesentery fat	Less	More
Peyer's Patches (Lymph Nodules)	Absent	Present in the lower portion
Vascularity	Greater	Lesser
Arteries	Jejunal branches of superior mesenteric	Ileal branches of the Superior mesenteric Ileocolic may supply the lower ileum
Arcades	Fewer number	Greater number
Arcade arteries	Straight and longer	Shorter

Inferior Mesenteric Artery

The inferior mesenteric artery originates approximately 4 inches above the aortic bifurcation at the L3 vertebral level. It runs downward and to the left across the aorta, psoas, and left common iliac to supply the 2nd half of the colon and end as the superior rectal artery.

The branches of the inferior mesenteric are:

I. **LEFT COLIC (287)—supplies the descending colon and the distal portion of the transverse colon**

   A. Ascending Branch—supplies the distal transverse and descending colon—**anastomoses** with the left branch of the middle colic near the splenic flexure

   Occlusive and nonocclusive vascular disease may cause ischemic colitis characterized by crampy left abdominal pain and bloody diarrhea. The splenic flexure is the most common site of this condition.

   B. Descending Branch—supplies the upper sigmoid colon—**anastomoses** with the ascending branches of the sigmoid arteries

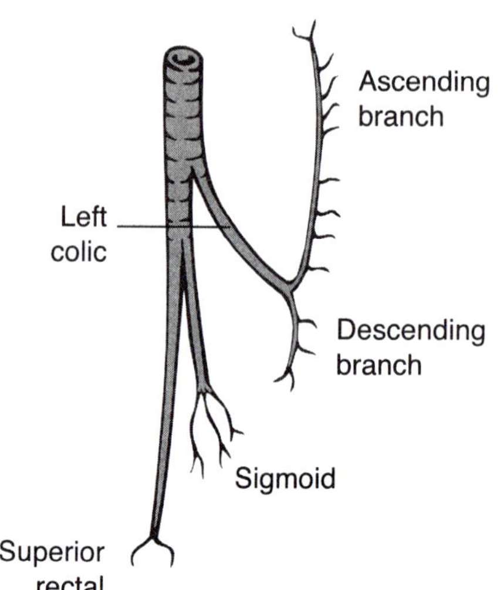

II. **SIGMOID ARTERIES (287)—go over the psoas to the sigmoid colon to the level of the rectum and upper anus**

   A. Marginal—in the pelvic mesocolon. The uppermost sigmoid artery has **anastomoses** with the descending branch of the left colic.

III. **SUPERIOR RECTAL (287)—supplies the rectum and the upper half of the anus**

Colon

I. COURSE—The colon is approximately 6 feet long. It begins as the cecum in the right iliac fossa, where it is covered on 3 sides by visceral peritoneum. The ascending colon continues superiorly to the hepatic (right colic) flexure, where it is related to the right kidney. The transverse colon continues slightly upward to the splenic (left colic) flexure, where it is related to the spleen and left kidney. The descending colon continues along the posterior abdominal wall in front of the kidney. In front of the quadratus lumborum and iliacus muscles, it becomes the sigmoid colon (anterior to the external iliac artery and at the medial aspect of the left psoas major muscle). The sigmoid colon becomes the rectum over the middle aspect (approximately the level of S3) of the sacrum.

➢ **II. SPECIAL FEATURES** (267)

A. Taeniae Coli—3 longitudinal muscular bands. The taeniae extend from the cecum to the rectum.

B. Haustra—sacculation of the wall (between taeniae)

C. Appendices Epiploicae—small peritoneal sacculations filled with fat

D. Caliber—decreases along its length

E. Flexures

1. *Hepatic—at the junction of the ascending and transverse colon*
2. *Splenic—at the junction of the transverse and descending colon—usually located higher than the hepatic flexure*

F. Mucosa—parallel rows of epithelial tubules (crypts) surrounded by lamina propria (connective tissue framework). The tubules are lined by goblet cells. The surface epithelium between tubules consists of columnar cells.

➢ **III. ARTERIES** (287)**—The proximal colon is supplied by the superior mesenteric; the distal colon by the inferior mesenteric.**

➢ **IV. VEINS** (292)**—correspond to arteries**

➢ **V. LYMPH NODES OF THE COLON** (297)

A. Local (Paracolic)—are located:

1. *Above the transverse colon and retroperitoneal*
2. *Medial to the ascending, descending, and iliac portions of the sigmoid colon and retroperitoneal*
3. *Along the mesentery of the pelvic portion of the sigmoid colon*

Chapter 5 *Abdomen*

B. Intermediate (Regional)—located along the colic arteries and named accordingly

C. Central (Main)—located near the origins of the superior and inferior mesenteric arteries

VI. NERVES (305, 306)

A. Proximal colon—superior mesenteric plexus and the Vagus

B. Distal colon—inferior mesenteric plexus and pelvic parasympathetics

VII. CECUM (264, 265)—the beginning of the colon—a blind pouch—frequently located in the right iliac fossa

A. Relationships—may vary

1. *Posterior—iliopsoas muscle, lateral femoral cutaneous nerve of the thigh, and external iliac artery*
2. *Posteromedial—the appendix*
3. *Superomedial—the ileocecal orifice and valve (a projection of mucous membrane into the cecal lumen)*

B. Special Features

1. *Ileocecal valve—at the junction of the ileum and cecum*
2. *Frenulum of the ileocecal valve—a horizontal mucosal ridge above the ileocecal valve*
3. *Cecal taeniae—converge at the base of the appendix to form its longitudinal muscle coat*

C. Folds

1. *Ileocolic—extends from the terminal ileum to the cecum—forms the anterior portion of the ileocecal recess*
2. *Retrocecal—from the cecum to the iliac fossa*

D. Fossae (Recesses)

1. *Retrocecal recess—a peritoneal reflection located behind the cecum. It may contain the appendix.*
2. *Superior ileocecal (Ileocolic) fossa—above the junction of the ileum and colon—formed by the ileocolic fold*
3. *Ileocecal fossa—formed by a peritoneal fold from the ileum to the cecum and appendiceal mesentery*

E. Arteries—cecal from the ileocolic from the superior mesenteric

F. Veins—ileocolics to superior mesenterics

G. Lymph Nodes—ileocolic and thence to the superior mesenteric nodes

H. Nerves—celiac and superior mesenteric ganglia—parasympathetics via vagal fibers

➢VIII. APPENDIX (264, 266)—originates about 1 inch below the ileocecal junction—contains lymphoid tissue. The mesoappendix is a peritoneal fold frequently connected to the ileal mesentery. The appendicular artery is usually located in the mesoappendix. It is a branch of the ileocolic artery.

A. Possible Locations of the Appendix

1. *Retrocecal (⅔ of appendices are retrocecal)*

2. *Retrocolic (alongside the colon)*

3. *Pelvic (the appendix extends inferiorly into the true pelvis)*

IX. ASCENDING COLON (254)

A. Relationships

1. *Anterior—omentum*

2. *Posterior—iliacus, quadratus lumborum, and transversus muscles and the right kidney*

B. Arteries—ileocolic and right colic from the superior mesenteric

C. Veins—correspond to arteries

D. Lymphatics—paracolics (located medial to the ascending colon) to intermediate and superior mesenteric nodes

E. Nerves—superior mesenteric plexus—parasympathetics via vagal fibers

X. TRANSVERSE COLON (254)—the largest and longest portion of the colon

A. Ligaments

1. *Phrenicocolic ligament—suspends left colic flexure to the abdominal wall (diaphragm)*

2. *Transverse mesocolon—attached to the pancreas*

B. Relationships

1. *Anterior—omentum*

2. *Posterior—the 2nd part of the duodenum, pancreas, jejunum, and ileum*

3. *Superior—stomach and gallbladder*

4. *Inferior—small intestine*

C. Arteries

1. *Middle colic—from the superior mesenteric—runs in the transverse mesocolon—supplies the proximal transverse colon*

    i. right branch

    ii. left branch

2. *Left colic—from the inferior mesenteric—supplies the distal transverse colon*

D. Veins—correspond to arteries

E. Lymphatics

1. *Proximal transverse colon—paracolics (located above the transverse colon) to intermediate and superior mesenteric nodes*

2. *Distal transverse colon—to inferior mesenteric nodes*

F. Nerves

1. *Proximal transverse colon—superior mesenteric plexus—parasympathetics via vagal fibers*

2. *Distal transverse colon—inferior mesenteric plexus*

XI. DESCENDING COLON (254)—extends from the left colic flexure to the iliac crest

A. Relationships

1. *Anterior—omentum*

2. *Posterior—left kidney; quadratus lumborum, iliacus, and psoas muscles; iliohypogastric, lateral cutaneous, ilioinguinal, and femoral nerves*

B. Arteries—left colic and sigmoid of the inferior mesenteric

C. Veins—corresponding (accompany the left colic artery)

D. Lymphatics—paracolic (located medial to the descending colon) to intermediate and inferior mesenteric nodes

E. Nerves—inferior mesenteric plexus—afferents via lumbar splanchnics (L1–L2)

XII. SIGMOID COLON (254)—extends from the iliac crest to the level of S3—is suspended by the pelvic (sigmoid) mesocolon

A. Portions

1. *Fixed—iliac portion—no mesentery*

2. *Mobile—pelvic portion—suspended by the pelvic mesocolon*

B. Arteries—sigmoids from the inferior mesenteric

C. Veins—correspond to arteries

D. Lymphatics—paracolics (located medial to the iliac portion and along the mesentery of the pelvic portion) to intermediate and inferior mesenteric nodes

E. Nerves—sympathetics via lumbar sympathetic trunk—parasympathetics via pelvic splanchnics

XIII. **ABNORMALITIES OF PERISTALSIS—regulated by the ANS and ENS. These may result in megacolon.**

A. **Hirschsprung's Disease—a congenital abnormality, predominantly of the distal colon, resulting from a derangement of the parasympathetic and enteric nervous systems (absence of VIP and NO relaxant ganglia). The functional defect is the inability of an aganglionic bowel segment to relax and allow a peristaltic wave to pass.**

B. **Crohn's Disease—a chronic inflammatory condition affecting the GI tract, especially the areas supplied by the mesenteric arteries**

C. **Chagas' Disease (Intestinal trypanosomiasis)—affects the colon and esophagus. The parasite produces a neurotoxin.**

XIV. COLONIC ABNORMALITIES

A. **Volvulus—An enlarged mesentery is responsible for a twisting of midgut derivatives, particularly the cecal and sigmoid portions of the colon. The resulting volvulus may also compromise the blood supply and produce a surgical emergency.**

B. **Intussusception—the infolding of a segment of bowel into an adjacent downstream segment. It is characterized by bright red bleeding or currant jelly stools in children. In adults it is frequently related to a malignant tumor.**

1. ***Ileocecal—the ileum infolds into the cecum***

2. ***Colocolic—a portion of the colon infolds into itself***

C. **Colonic polyps are inward mucosal growths characterized by bleeding. They are prone to become malignant.**

D. **Diverticula—small outpocketings of the wall of the colon. Like the appendix, they are subject to inflammation (diverticulitis) and perforation.**

TABLE 5-7. SMALL AND LARGE INTESTINES

PART	SMALL INTESTINE	LARGE INTESTINE
Length	Duodenum—9 inches Jejunum and Ileum—20 feet	6 feet
Caliber	Duodenum—4 centimeters	Decreases along its length (8 to 3 centimeters)
Mucosa—Epithelium	Simple columnar—enterocytes and goblet cells—small tubular glands	Parallel rows of epithelial crypts lined by goblet cells and surrounded by the lamina propria Surface epithelium between crypts—columnar cells
Submucosa	Duodenum—Brunner's glands Jejunum—Brunner's glands absent Ileum—lymph nodules (Peyer's patches)	Connective tissue contains more blood vessels and lymphatics
Villi	Present	Absent
Muscularis	Outer—Longitudinal layer Inner—Circular layer	Outer—Taeniae Coli Inner—Circular layer
Plicae	Present	Absent
Mesentery	Jejunum and Ileum	Cecum, Transverse and Sigmoid Colon
Ligaments	Duodenum—Ligament of Treitz	Transverse colon Phrenicocolic Transverse mesocolon
Taeniae	Absent	Present
Appendices Epiploicae	Absent	Present
Haustra	Absent	Present—Absent in Cecum and Sigmoid
Openings	Ampulla of Vater	Appendix
Diverticula	Meckel's (occasional)	Diverticula occasionally present
Arteries	Celiac—Foregut Portion Superior Mesenteric—Midgut Portion	Proximal colon—Superior Mesenteric Distal colon—Inferior Mesenteric
Veins	Pancreaticoduodenal and Superior Mesenteric	Colics to mesenterics
Lymphoid Tissue	Ileum—Peyer's Patches (Lymph nodules)	Appendix contains lymph nodules
Nerves	Celiac and Superior Mesenteric Plexuses, Vagus	Proximal colon—Superior Mesenteric Plexus, Vagus Distal colon—Inferior Mesenteric Plexus and Pelvic Parasympathetics

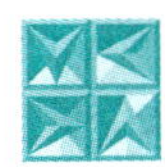

Kidneys

The kidneys are retroperitoneal organs surrounded by fat.

I. STRUCTURE (311, 313, 317, 318)

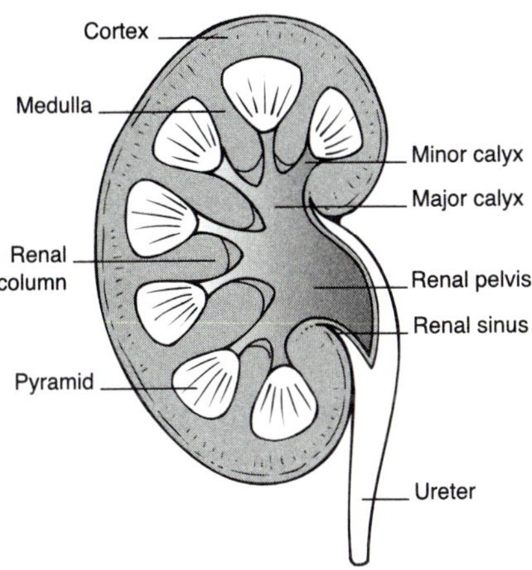

A. Capsule of Gerota (Tunica Fibrosa)—outer fibrous tissue on the surface of the kidney. The capsule is surrounded by perirenal fascia (a continuation of the transversalis fascia).

B. Cortex—contains glomeruli and convoluted portions of renal tubules—covers the bases of the renal pyramids

1. *Corpuscle*
   i. glomerulus—an arterial capillary net (afferent and efferent glomerular arteries)
   ii. loops of Henle—located in the cortex **and** medulla
2. *Bowman's capsule—an epithelial (simple squamous) capsule surrounding the corpuscle. The capsule opens into a tubule and thence into a renal sinus.*

C. Medulla—consists of pyramids—contains ascending and descending limbs of renal tubules (**loops of Henle**) and collecting tubules with their respective arteries

1. *Loop of Henle—contains ascending and descending limbs of straight tubules but* ***not*** *the collecting tubules*

D. Renal Sinus—located at the concave border of the kidney

E. Renal Hilus—the opening of the renal sinus and the entrance for the renal vessels—contains the renal pelvis and calyces

F. Renal Pelvis—formed by the renal calyces and the expanded end of the ureter. Renal vessels lie anterior to the renal pelvis.

G. Calyces—Major (2 or 3)—Minor (approximately 8)

H. Kidney Lobe—a renal pyramid and its associated cortical portion

I. Segments—superior, inferior, anterosuperior, anteroinferior, and posterior

J. Renal Pedicle—the proximal ureter and the renal vessels, nerves, and lymphatics

II. RELATIONSHIPS (312, 324)

A. Right Kidney (is lower than the left)

1. *Anterior—suprarenal, liver, duodenum, and colon*
2. *Posterior—ribs 11 and 12—diaphragm, psoas, quadratus lumborum, and transversus abdominis muscles—iliohypogastric and ilioinguinal nerves*

B. Left Kidney

1. *Anterior—suprarenal, spleen, stomach, pancreas, colon, and part of jejunum*

2. *Posterior—ribs 11 and 12—diaphragm, psoas, quadratus lumborum, and transversus abdominis muscles—iliohypogastric and ilioinguinal nerves*

➢ **III. RENAL ARTERY** (314–316)**—a branch of the aorta. It lies behind the renal vein. Before entering into the kidney, the renal artery forms anterior and posterior branches. The right renal artery is longer than the left.**

A. Segmental Arteries—5 (or more) go to renal segments

1. *Lobar arteries—one to each pyramid*

i. interlobar—run alongside the pyramids toward the cortex

a. arcuate—arch over bases of pyramids

1) interlobular—into the cortex to the afferent glomerular

i) afferent glomerular—supply glomeruli

ii) efferent glomerular—leave glomeruli to descend into the medulla to supply those loops of Henle which are located in the medulla

➢ **IV. RENAL VEIN** (314, 316)**—lies anterior to the renal artery—a tributary of the inferior vena cava. The left renal is longer than the right. The left ovarian/testicular vein empties into the left renal. The suprarenals and inferior phrenics empty into the renal vein.**

➢ **V. LYMPHATICS** (321)**—located behind the renal pelvis. They drain into aortic lymph nodes.**

➢ **VI. NERVES** (322, 323)

A. Visceral Afferent Sensory fibers travel via sympathetic fibers in the renal plexus and thence to lesser and least (lowest) splanchnic nerves to T11 to L1.

Renal pain may be referred to the lumbar and inguinal regions.

B. Sympathetic—from the lesser and least (lowest) splanchnic nerves to the renal plexus. They constrict vessels and decrease urine production. Postganglionic fibers from the renal plexus go to the kidney, suprarenal, and upper ureter.

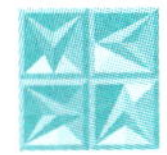

Ureters

The ureters are 10-inch-long muscular tubes which transport urine from the kidneys to the bladder.

I. COURSE (319, 320)—**A ureter begins in the hilus of the kidney and extends downward retroperitoneally along the psoas muscle. It crosses the common iliac in front of the sacroiliac joint and then goes over the pelvic brim** (345) **(*a surgical danger area because it is close to the iliac artery*). The ureter continues down the lateral pelvic wall and proceeds under the uterine artery** (320) **(*a danger area*) and into the bladder** (320) **(*a danger area*).**

II. RELATIONSHIPS (319, 320)

A. Right Ureter

1. *Anterior—duodenum, ileum, and testicular or ovarian vessels*
2. *Posterior—psoas muscle and bifurcation of right common iliac*

B. Left Ureter

1. *Anterior—colon and testicular or ovarian vessels*
2. *Posterior—psoas muscle and bifurcation of left common iliac*

III. ARTERIES (320)**—along its course**

A. Renal—supplies the upper ureter

B. Testicular or Ovarian—supplies the middle of the ureter

C. Superior Vesical—supplies the lower ureter. Near the bladder, the superior vesical artery lies above the ureter.

D. Inferior Vesical (male) or Uterine (female)—may supply the portion of the ureter near the bladder. The middle rectal, common iliac, and internal iliac may send branches to the ureter.

IV. VEINS (320)**—correspond to arteries**

V. LYMPHATICS (321)

A. Upper Ureter—renal or aortic

B. Middle Ureter—common iliac

C. Lower Ureter—external or internal iliac

VI. NERVES (322, 323)

A. Visceral Afferent Pain fibers from the upper ureter travel via renal plexus to vertebral levels T12-L1—pain from the midureter to L1–L2. Therefore, pain can be referred to the lumbar and inguinal regions and the thigh.

B. Sympathetic—via renal or ureteric plexus to the superior hypogastric plexus

Suprarenals

I. GENERAL—The suprarenals (adrenals) (325) are hormone-secreting endocrine glands located on the upper poles of the kidneys.

A. Cortex—secretes minerals and glucocorticoids (androgens and estrogens)—mesodermal origin. The zona fasciculata and zona reticularis are stimulated by ACTH.

B. Medulla—secretes catecholamines and dopamine—neural crest (ectodermal) origin

II. RELATIONSHIPS (325)

A. Right Suprarenal—lies behind the right lobe of the liver and the inferior vena cava

1. Superior—the crus of the diaphragm

2. Anterior—the liver and inferior vena cava

3. Posterior and inferior—the kidney (connective tissue lies between the kidney and suprarenal)

B. Left Suprarenal—lies behind the pancreas and stomach

1. Superior—the spleen

2. Anterior—the stomach

3. Posterior—the crus of the diaphragm and the kidney

4. Inferior—the pancreas

III. SUPRARENAL ARTERIES (325)

A. Superior—from the inferior phrenic

B. Middle—from the aorta

C. Inferior—from the renal

IV. SUPRARENAL VEINS (325)

A. Right—drains into the posterior surface of the inferior vena cava

B. Left—drains into the renal vein

V. LYMPHATICS (321)—to aortic nodes

VI. SYMPATHETIC NERVES (326)

A. The Medulla receives **preganglionic** fibers from the celiac plexus. The chromaffin cells increase production of epinephrine and norepinephrine. The cortex does not have innervation.

B. Blood Vessels in the suprarenals receive postganglionic fibers from the celiac ganglia.

TABLE 5-8. LUMBAR PLEXUS (463, 464)

NERVE	COURSE	IN THE LOWER ABDOMEN, PERINEUM AND THIGH SUPPLIES
Femoral (L2–L4)	Emerges in the psoas (between the psoas and iliacus)	Skin of the anterior thigh Iliopsoas, muscles of the anterior thigh, and other muscles of the lower limb
Genitofemoral (L1–L2) Genital Femoral	Pierces the psoas muscle → posterior to the inguinal ligament	Skin of scrotum and medial thigh Cremaster muscle Skin of the anterior thigh
Lateral (Femoral) Cutaneous Nerve of the Thigh (L2–L3)	Pierces psoas → under the inguinal ligament	Skin of the anterolateral thigh
Iliohypogastric (L1)	Lateral abdominal wall to the lower abdominal wall	Anterior abdominal wall muscles Skin of the lower abdomen
Ilioinguinal (L1)	Anterior to the quadratus lumborum → abdominal muscles → inguinal canal	Skin of the lower abdomen, scrotum, and labia majora Muscles of the lower abdominal wall
Obturator (L2–L4)	Pierces psoas → laterally → obturator foramen	Part of the skin of the medial thigh Adductor muscles of the thigh

TABLE 5-9. SACRAL PLEXUS (463, 465)

NERVE	COURSE	IN THE PERINEUM OR LOWER LIMB SUPPLIES
Sciatic (L4-S3)	Thru the greater sciatic foramen → between the greater trochanter and ischial tuberosity → divides into the tibial and common peroneal (fibular) in the thigh	Most of the lower limb
Posterior (Femoral) Cutaneous Nerve (S1–S3)	Thru the greater sciatic foramen	Skin of the lower buttock, posterior thigh, and posterior upper leg
Pudendal (S2–S4)	Pelvis → thru the greater sciatic foramen → thru the lesser sciatic foramen	Perineum
Nerves to muscles		Levator ani, gluteals, piriformis, quadratus femoris, and obturator internus

Pelvis

Page numbers in black. Netter plate numbers in green.

Divisions 354
Supply 364, 382
Pelvic diaphragm 333–337
Urogenital diaphragm 352, 357

PERINEUM—MALE 354–356 213

Urogenital region 338
Anal region 354, 364

PERINEUM—FEMALE 217

Urogenital region 351, 352
Anal region 364
Perineal body (central tendon, perineal center, central tendinous point) 351, 375

RECTUM 221

Location 366
Peritoneal relationships 368
Folds 365
Muscles 366
Arteries 369
Veins 370
Lymphatics 297
Nerves 306, 383

ANUS 222

Pectinate (dentate) line 365
Anal muscles 366, 367
Spaces 368
Other lines 365

Bony Pelvis

I. FOUR BONES (330, 331)

A. Sacrum—large triangular bone at the lower part of the vertebral column

1. *Sacral promontory*
2. *Sacral hiatus (due to incomplete closure of the arch of L5)—**site of insertion of the needle for caudal anesthesia***
3. *Sacral canal contents: last 5 sacral nerves, lateral sacral arteries, and filum terminale*

B. Coccyx—formed from 4 rudimentary vertebrae

C. Innominates (2) (Coxals)

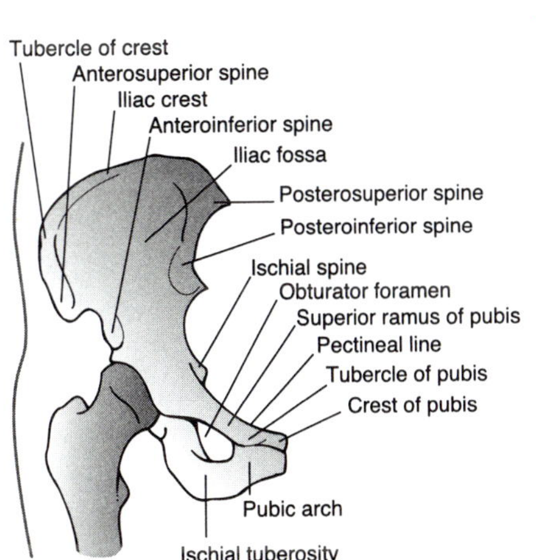

1. *Ilium—the superior portion of the innominate bone extending superiorly from the acetabulum. The upper portion forms alae and the inferior portion contributes to the acetabulum.*
   - i. iliac spines (4)—anterior and posterior. (Each has a superior and an inferior.)
     - a. anterosuperior iliac spine—an attachment for the inguinal ligament—located at the junction of the iliac crest and the anterior border of the ala of the ilium
     - b. anteroinferior iliac spine—on the anterior surface of the ala and below the notch
   - ii. iliopectineal eminence—located midway between the pubic symphysis and the sacroiliac articulation
   - iii. the crest of the ilium—attachment for abdominal and back muscles—lies at the level of the L4 vertebra and the bifurcation of the aorta
2. *Ischium—body, tuberosity, and 2 rami (inferior rami). The obturator foramen is located between the ischium and the pubis.*
   - i. body—for the attachment of ischiocavernosus and transverse perineal muscles
   - ii. ramus (inferior ramus)—joins the inferior ramus of the pubis
   - iii. ischial spine—attachment for muscles (levator ani and gemellus superior) and the sacrospinous ligament

     The character of the spines can be an important factor in obstetrics (descent of the fetal head).
   - iv. ischial tuberosity—located posteriorly on the body

Chapter 6
Pelvis

3. *Pubis—the anterior part of the innominate bone*
   i. body—forms part of the acetabulum
   ii. rami
      a. superior ramus—extends from the body to the median plane—joins the ischium and ilium in the acetabulum
      b. inferior ramus—extends laterally and downward from the medial end of the superior ramus
   iii. pubic symphysis—articulation between pubic bones
   iv. pubic tubercle—the medial attachment of the inguinal ligament

II. TRUE PELVIS (Pelvis Minor) (332)—Boundaries:

A. Inlet—Linea terminalis (Terminal line) (Pelvic brim)
   1. *The promontory of the sacrum and anterior border of sacral alae*
   2. *Iliopectineal line—a ridge on the ilium and pubis delineating a portion of the brim of the true pelvis*
      i. iliac portion—of the iliopectineal line—sometimes called the arcuate line (not to be confused with the arcuate line of the posterior rectus sheath, the arcuate eminence of the temporal bone, the arcuate ligament of the diaphragm, or the arcuate ligament of the urogenital diaphragm)
      ii. pectineal portion—of the pectineal line (pectin pubis) (not to be confused with the pectinate line of the anus)

B. Sides of the True Pelvis
   1. *Sacrum and coccyx*
   2. *Symphysis pubis—the articulation between the pubic bones*
   3. *Ischia and part of ilia*
   4. *Sacrospinous (Anterior sacrosciatic) ligaments—from the sacrum to the ischial spines*

C. Outlet
   1. *Sacrotuberous ligaments—from the sacrum to the posteroinferior iliac spines*
   2. *Tip of coccyx*
   3. *Lower borders of pubic bones (inferior pubic rami) and symphysis*
   4. *Ischial tuberosities*

III. FALSE PELVIS—lies above the pelvic inlet. Its boundaries are:

A. Anterior—lower abdominal wall

B. Lateral—iliac bone

C. Posterior—alae of the sacrum and L5 and S1 vertebrae

IV. ARTICULATIONS (331)

A. Sacroiliac—between the articular surfaces of the sacrum and ilium. It contains a layer of cartilage.

B. Symphysis Pubis—joint between 2 oval articular surfaces. It is supported by the superior and arcuate pubic ligaments.

V. LIGAMENTS (330, 331)

A. Sacrospinous—from the caudal border of the sacrum to the ischial spine

B. Sacrotuberous—from the lateral aspect of the sacrum and coccyx to the ischial tuberosity

C. Sacroiliac

D. Sacrococcygeal—**may be injured during childbirth and be painful post partum**

E. Inguinal—the aponeurosis of the external oblique

1. *Lacunar (Gimbernat's)—that portion of the inguinal ligament which is reflected downward and backward*

2. *Pectineal (Cooper's)—a continuation of the inguinal and lacunar ligaments onto the pubic periosteum—covers the pectineal line*

F. Sacrolumbar—L5 to the sacrum

G. Iliolumbar—S4 and S5 to the iliac crest

H. Arcuate—between the pubic rami—forms the upper portion of the pubic arch

VI. FORAMINA (330, 331)

A. Greater Sciatic—formed by sciatic ligaments (sacrospinous and sacrotuberous)

B. Lesser Sciatic—formed by sciatic ligaments

C. Obturator—large and fibrous with a small opening (obturator canal) which transmits the obturator vessels and nerve

A herniation thru the obturator foramen may involve the small bowel and result in bowel obstruction. It may be diagnosed by noting a tender mass on pelvic exam and pain in the anteromedial thigh due to pressure on the obturator nerve.

D. Sacral—transmit the anterior primary rami of the sacral spinal nerves

VII. PELVIC PLANES (332)

A. Plane of the Inlet

1. *AP diameter—from the posterior surface of the symphysis to the promontory of the sacrum*
2. *Left oblique diameter—from the left sacroiliac articulation to the right iliopectineal eminence*
3. *Right oblique diameter—from the right sacroiliac eminence to the left iliopectineal eminence—longer than the left*
4. *Transverse diameter—**The fetal head enters the widest diameter of the inlet of the pelvis.***

B. Plane of Greatest Pelvic Dimension

1. *AP diameter—from the middle of the pubic symphysis to the junction of S2 and S3*
2. *The distance between the lateral surfaces of the pelvis*

C. Plane of Least Pelvic Dimension—the distance between the iliac spines (bispinous distance)

A decrease in this midplane diameter may result in an arrest in descent of the fetal head during labor.

D. Planes of the Outlet—The outlet is formed by two triangles. The base of each triangle is the distance between the ischial tuberosities.

1. *Anterior triangle*
   i. base—between the ischial tuberosities
   ii. sides—rami
   iii. apex—at the inferior border of the symphysis
2. *Posterior triangle*
   i. base—between the ischial tuberosities
   ii. sides—sacrotuberous ligaments
   iii. apex—tip of the coccyx

VIII. CLINICAL PELVIMETRY (332)—defined by pelvic diameters

A. AP of the Inlet—from the posterior surface of the symphysis to the promontory of the sacrum

B. True Conjugate—from the promontory of the sacrum to the upper border of the pubic symphysis

C. Obstetric Conjugate—from the promontory of the sacrum to a point on the pubic symphysis just below its upper margin

D. Diagonal Conjugate—from the promontory of the sacrum to the inferior border of the pubic symphysis—**clinically determined by vaginal examination**

E. Interspinous—between the anterosuperior iliac spines

F. Bispinous—between the ischial spines

G. Transverse of the Inlet—the widest inlet diameter (13–14 cm)

H. Transverse of the Outlet—between the inner aspects of the ischial tuberosities—**measured with an instrument or the fist**

I. AP of the Outlet—from the inferior border of the pelvic symphysis to the tip of the coccyx

IX. MALE AND FEMALE PELVIS, DIFFERENCES (332)

TABLE 6-1.

STRUCTURE	MALE	FEMALE
Bone weight	Heavier	Lighter
Iliac bones	Vertical tendency	Flare outward
Sacrum	Straighter	More curved
Sacral length	Longer	Shorter
Sacral inclination	More forward	Less forward
Sacrosciatic notch	Narrower	Wider
Ischial spines	Sharper	More blunt
Intertuberous and Interspinous diameters	Narrower	Wider
Sidewalls	Converging	Diverging
True pelvis	Narrower and deeper	Shallower and wider
Pelvic inlet	More heart shaped	Rounder
Pubic symphysis	Narrower	Wider
Subpubic arch	Narrower	Wider
Subpubic angle	More acute	Less acute
Obturator foramen	More oval	More triangular

Pelvic and Perineal Muscles

I. MUSCLES WHICH PASS OUT OF THE PELVIS TO INSERT ON THE FEMUR (461, 462)

A. Iliacus (L2–L4)

1. *Passes beneath the inguinal ligament*
2. *From the ilium to the lesser trochanter—flexes the thigh*

B. Psoas Major (L2–L4)

1. *Passes beneath the inguinal ligament*
2. *From the lumbar vertebrae to the lesser trochanter—flexes the thigh*

C. Obturator Internus (L5-S1)—anterolateral covering of the true pelvis

1. *Passes thru the lesser sciatic foramen*
2. *From the obturator membrane to the greater trochanter—abducts the thigh and steadies the femoral head*

D. Piriformis (S1–S2)

1. *Passes thru the greater sciatic foramen*
2. *Posterior covering of the true pelvis*
3. *From the sacrum to the greater trochanter of the femur—abducts the thigh—rotates the femur outward*

II. PELVIC DIAPHRAGM (333–338, 343–345)—muscular support for pelvic viscera

A. Levator Ani (S3–S4)—portions insert on the central tendon (perineal body)—maintains the integrity of the pelvic floor—supports the viscera and contributes to urinary and fecal continence

1. *Pubococcygeus—from the pubis to the coccyx*
2. *Iliococcygeus—from the obturator fascia and ischial spine to the coccyx*
3. *Puborectalis—constricts the rectum and vagina—forms a "rectal sling"*

B. Coccygeus (S4–S5)—from the ischial spine to the lateral margin of the lower sacral and upper coccygeal vertebrae. The coccygeus overlies the sacrospinous ligament and is inferior to the piriformis muscle.

III. UROGENITAL DIAPHRAGM (352) (Deep Perineal Compartment) (Pouch) (Space)—composed of muscle and 2 layers of fascia. It is a strong musculomembranous partition stretched almost horizontally across the anterior half of the lower pelvis and below the pelvic diaphragm. The UG diaphragm contains the bulbourethral glands.

A. Urethral Sphincter

1. *The external urethral (voluntary) sphincter—a continuation of the deep transverse perineal muscle*
2. *From the ischial ramus to the opposite ischial ramus (around the urethra)*

B. Deep Transverse Perineal muscle—from the ischial rami to the lateral aspects of the vagina (female) or to the opposite ischial ramus (male)

IV. SUPERFICIAL PERINEAL MUSCLES (352, 355)

A. Superficial Transverse Perineal—from the ischial tuberosities to the central tendon (perineal body)

B. External Anal Sphincter (voluntary)

C. Male

1. *Bulbospongiosus (Bulbocavernosus)—from the central tendon to the penis*
2. *Ischiocavernosus—from the ischium to the crura of the penis (contraction helps maintain erection)*
3. *Cremaster—a continuation of the internal oblique—elevates the testes*

   NOTE: *The genital branch (motor) of the genitofemoral nerve (L1–L2) innervates the cremaster muscle. The femoral branch (sensory) of the genitofemoral nerve innervates the medial portion of the thigh.*

   These innervations form the basis of the "cremasteric reflex."

D. Female

1. *Bulbospongiosus—from the central tendon, around the vaginal opening (covering the corpora cavernosa), to the body of the clitoris*
2. *Ischiocavernosus—from the medial portion of the ischial rami to the clitoral crura—(constriction helps maintain clitoral erection)*

Pelvic Arteries

The main arteries supplying pelvic structures are the ovarian, inferior mesenteric, and internal iliac.

I. OVARIAN (371)—arises from the aorta just below the origin of the renal to travel retroperitoneally across the psoas and then ventral to the ureter and the common iliac into the suspensory ligament of the ovary (infundibulopelvic) to the hilum of the ovary. It supplies the tube and ovary and has anastomoses with the uterine artery.

II. INFERIOR MESENTERIC (369)—arises from the aorta 3 centimeters above its bifurcation. A branch of the inferior mesenteric (left colic) in the abdomen supplies the descending colon.

A. Sigmoidal (287)—to the sigmoid colon

B. Superior Rectal (287)—the continuation of the inferior mesenteric into the pelvis—has **anastomoses** with the middle rectal, a branch of the anterior division of the internal iliac, and inferior rectal, a branch of the internal pudendal

III. COMMON ILIAC (369)—formed by bifurcation of aorta at the L4 vertebral level. It divides lateral to the sacroiliac ligament.

A. External Iliac (236)—extends along the medial border of the psoas to pass beneath the inguinal ligament and become the femoral artery

1. Inferior epigastric (238)—2 small branches are the pubic and cremasteric

2. Deep circumflex iliac (236, 238)—courses between the internal oblique and transversus to supply a portion of the anterior abdominal wall.

B. Internal Iliac (Hypogastric) (373)—The major artery to the pelvis. It divides into anterior and posterior divisions near the greater sciatic foramen.

1. Posterior division

i. iliolumbar (247)—to iliac muscles and bone, and the lumbar area

ii. lateral sacral (247)—thru the sacral foramina to supply skin over the sacrum

iii. superior gluteal (247, 320, 468)—thru the greater sciatic foramen to the gluteal region

2. *Anterior division*

i. inferior gluteal (247, 373, 468)—thru the greater sciatic foramen to the gluteal region and posterior thigh

ii. obturator (373)—thru the obturator canal

iii. internal pudendal (376)—thru the greater sciatic foramen over the ischial spine to return to the pelvis thru the lesser sciatic foramen. In the pelvis it runs in the pudendal canal of Alcock, located in a fat-filled space (between the levator and the obturator internus) called the **ischioanal (ischiorectal) fossa.** The pudendal artery ends by dividing into the perineal artery and the artery to the penis/clitoris.

a. inferior rectal (hemorrhoidal) (369)—supplies the levator ani muscle, anus, and perineal skin.

b. perineal (375)—has scrotal/labial branches

c. arteries to the penis or clitoris (376)

1) dorsal artery to the penis (355, 357, 374) or clitoris

2) deep artery (357)—extends thru the UG diaphragm to go to the corpus cavernosum and corpus spongiosum (important in erection)

3) urethral—to the penile urethra

iv. superior vesical (371)—to the bladder

v. inferior vesical (371)—to the bladder and prostate

vi. umbilical (217)—(obliterated) dense fibrous tissue called the umbilical ligament

vii. middle rectal (369)—to the lower rectum and vagina

viii. uterine (344, 345, 371)—gives off a cervical branch and then ascends upward alongside the cervix, where it passes superior to the ureter, then continues upward sending branches to the uterine muscle, and thence proceeds further upward to **anastomose** with the ovarian artery

a. vaginal—supplies the upper vagina and the fundus of the bladder. The vaginal artery may be a branch of the anterior division of the internal iliac or a branch of the cervical. It is comparable to the inferior vesical artery in the male.

b. cervical

c. branches to the uterus

Pelvic Veins

The major pelvic veins generally follow the course of the arteries except for the left ovarian, which drains into the left renal, and the deep dorsal vein of the clitoris, which drains into the vesical venous plexus.

I. **INFERIOR VENA CAVA** (371–372)—**formed by the common iliacs at the level of L5 and just to the right of the median line**

   A. Right Ovarian and Testicular (the left empties into the renal)

II. **COMMON ILIAC VEINS** (371–374)—**formed, opposite the sacroiliac joint, by the external and internal iliacs. The common iliac veins pass behind the common iliac arteries.**

   A. Middle Sacral Vein—communicates with the lateral sacral veins to form the sacral venous plexus which drains into the common iliac vein

III. **EXTERNAL ILIAC VEINS** (248, 373, 374)—**begin at the inguinal ligament. The femoral vein becomes the external iliac at the inguinal ligament.**

   Pressure from a pelvic mass, such as a pregnant uterus, may impede venous flow and produce saphenous vein dilation resulting in valvular incompetence and varicose veins.

   A. The external iliac veins receive the inferior epigastric and deep circumflex veins from the anterior abdominal wall.

IV. **INTERNAL ILIAC VEINS** (248, 292, 374)—**begin above the greater sciatic foramen**

   A. Superior and Inferior Gluteal Veins (292)

      1. *Drain the gluteus maximus and surrounding muscles*

      2. *Pass thru the greater sciatic foramen*

   B. Iliolumbar Veins (248)

   C. Lateral Sacral Veins (248)—veins on each side are connected to the sacral plexus

   D. Obturator Veins (248)—begin in the thigh and enter the pelvis thru the obturator foramina

   E. Internal Pudendal Veins (370, 374, 376)—drain the deep veins of the penis and prostatic plexus which drain the corpora cavernosa

   F. Uterine Veins—located in the broad ligament lateral to the uterus

V. PLEXUSES (248, 338, 343, 361, 370)—**Venous plexuses are networks of veins which connect with other veins. Posterior venous plexuses drain into the inferior mesenteric while the anterior venous plexuses drain into the internal iliac.**

Impediments to the free flow of blood within their lumina produce venous dilation and form the basis of the "pelvic congestion syndromes."

Pelvic plexuses are particularly susceptible to injury during surgical procedures.

A. Vesical (female) Plexus—surrounds the upper urethra and bladder neck. It drains into the internal iliacs.

   1. *Dorsal vein of the clitoris*

B. Vesical (male) Plexus—receives the prostatic plexus

   1. *Dorsal vein of the penis—has communications with the deep vein of the penis. The dorsal vein of the penis goes under the pubic symphysis.*

C. Prostatic Plexus—located anterolaterally to the prostate. It drains into the vesical plexus, the vertebral plexus, and the internal iliac veins.

D. Pampiniform (male) Plexuses—located in the inguinal canals. They drain the testes.

E. Rectal—The superior rectal vein (of the portal system) connects with the middle and inferior rectal veins (systemic veins).

   External or internal pressure on these veins may contribute to the formation of hemorrhoids because portosystemic anastomoses are located between the upper and lower portions of the anal canal.

F. Sacral Venous Plexus—located on the anterior surface of the sacrum. It communicates with the lateral and middle sacral veins and drains into the common iliac veins.

G. Ovarian Venous Plexus—located below the ovary and tube, between the two layers of the broad ligament. The ovarian vein communicates with the uterine vein.

Pelvic and Perineal Lymph Nodes

➢ I. **VULVA (377, 378)—to the superficial inguinal nodes to the deep inguinals to the iliacs**

➢ II. **CLITORIS (377, 378)—deep nodes inguinal and a few lymphatics to external iliacs via the inguinal canal**

➢ III. **VAGINA (377, 378)**

A. Lower ⅓—superficial inguinal

B. Middle ⅓—iliacs

C. Upper ⅓—iliacs and sacrals

➢ IV. **UTERUS (377, 378)—iliacs**

A. The Fundus—iliacs and some drainage to the aortics via ovarian lymphatics

B. Body—internal iliacs

C. Cervix—iliacs, rectal, and sacral

➢ V. **TUBES AND OVARIES (377, 378)—aortics and internal iliacs**

➢ VI. **FEMALE URETHRA (377, 378)—iliacs and sacrals**

➢ VII. **TESTES (379)—lumbar and aortics**

➢ VIII. **SEMINAL VESICLES (379)—iliacs**

➢ IX. **PROSTATE (379)—iliacs, obturator, and sacrals**

➢ X. **MALE URETHRA (379)—external iliacs**

➢ XI. **PENIS (379)—superficial inguinals**

A. Glans—deep inguinal, external iliacs

➢ XII. **SCROTUM (379)—superficial inguinals**

➢ XIII. **BLADDER (379)—iliacs**

➢ XIV. **RECTUM (297)**

A. Upper ½—inferior mesenterics

B. Lower ½—sacrals, superficial inguinal, and internal iliacs

➢ XV. **ANAL CANAL (297)—sacral, internal iliac, and superficial inguinals**

A. Above the pectinate line—sacral and internal iliacs

B. Below the pectinate line—superficial inguinal

Pelvic Nerves

I. FIBERS OF THE SOMATIC NERVOUS SYSTEM (380)

A. Lumbar Plexus (L1–L4) (464)

1. *Supplies muscles passing thru the pelvis from the abdomen to the thigh*
   - i. femoral (L2–L4)—muscular and cutaneous
   - ii. obturator (L2–L4)—leaves thru the obturator foramen to supply the medial thigh muscles and overlying skin
2. *Supplies cremaster muscle—genital branch of genitofemoral (L1–L2)—via inguinal canal*
3. *Supplies skin of lower abdomen, thigh, labia, and scrotum*
   - i. iliohypogastric (L1)—lateral thigh and pelvis
   - ii. ilioinguinal (L1)—medial thigh, labia, and scrotum
4. *Supplies skin of lateral thigh*
   - i. lateral cutaneous (L2–L3)
   - ii. femoral branch of the genitofemoral (L1–L2)

B. Lumbosacral Trunk (L4–L5) (463)—Fibers contribute to formation of the sacral plexus in the pelvis.

C. Sacral Plexus (L4-S4) (465)—12 primary nerves which supply the lower limbs, pelvic muscles, and perineum. The nerves of the sacral plexus unite to form the sciatic nerve. The primary perineal nerve is the pudendal nerve.

1. *Branches to the lower limb which exit the pelvis via the greater sciatic foramen*
   - i. to muscles
     - a. sciatic (L4-S3)—the largest
     - b. gluteals (L4-S1)
     - c. quadratus femoris (L4-S1)
     - d. obturator internus (L5-S1)
   - ii. to skin—posterior femoral cutaneous nerve of the thigh
2. *Branch to the piriformis (S2)*
3. *Branch to the levator ani (pelvic diaphragm) (S3–S4)*
4. *Pudendal (S2–S4)—exits pelvis via the greater sciatic foramen and enters the perineum via the lesser sciatic foramen. The pudendal nerve has motor, sensory, and sympathetic components.*

i. branches of the pudendal nerve

a. inferior rectal—to external rectal sphincter

b. perineal—to muscles of the UG diaphragm

c. dorsal nerve of the penis/clitoris

d. scrotal/labial

Pudendal block will not relieve labor (uterine) pain but will allow painless episiotomy.

II. FIBERS OF THE ANS IN THE PELVIS (381–383, 385–388)

A. Plexuses (collections of nerves)

1. *Superior hypogastric plexus—located below the bifurcation of the aorta*

   i. mostly sympathetic

   ii. divides into hypogastric nerves which contribute to the inferior hypogastric plexus

2. *Inferior hypogastric plexus*

   i. sympathetic and parasympathetic fibers

   ii. receives contributions from hypogastric nerves and sacral parasympathetics

   iii. fibers follow blood vessels to organs

B. Sympathetic (SNS) (T5-L2)

1. *Trunks*

   i. lumbar sympathetic trunk

   ii. sacral sympathetic trunk

   a. a continuation of the lumbar sympathetic trunk

   b. Terminal ends join to form the *ganglion impar.*

2. *Linkages*

   i. Preganglionic neurons originate in the spinal cord.

   ii. Postganglionic neurons originate in the hypogastric plexus or the lumbar plexus or in the lumbar or sacral sympathetic trunk and thence to pelvic viscera.

3. *Major organs supplied:*

   i. bladder (L1–L2)—SNS relaxes detrusor

   ii. urethra (L1–L2)—SNS contracts sphincter

iii. uterus (T11-L1)—SNS inhibits contractions

iv. ovaries and tubes (T5-L2)

v. ejaculatory ducts, seminal vesicles, prostate—SNS constrict and result in ejaculation

C. Parasympathetic (PNS)

1. *Fibers connect to the spinal cord via spinal nerves S2–S4.*

2. *Pelvic splanchnics—parasympathetics in the pelvis going to pelvic viscera*

3. *Linkages*

i. Preganglionic neurons originate in the sacral spinal cord.

ii. Postganglionic neurons originate in the inferior hypogastric plexus or in the walls of the viscera.

4. *Major organs supplied*

i. bladder wall—PNS contracts detrusor muscle

ii. sphincter urethrae—PNS relaxes sphincter

iii. penis and clitoris—PNS dilates blood vessels and enhances erection

iv. bulbourethral glands (male)—PNS increase secretions

v. vestibular glands (female)—PNS increase secretions

III. VISCERAL AFFERENTS IN THE PELVIS (386, 387)

A. **Pelvic visceral pain is brought about by organs which are inflamed, distended, or ischemic. It often consists of a deep, ill-defined ache.**

1. *Female organs*

i. uterus body—via inferior hypogastrics to T10–T11

ii. cervix—to S2–S4

iii. ovaries—via ovarian plexus to the preaortic and renal plexuses to T10–T11

iv. tubes—via ovarian plexus to T11-L1

v. vagina—said to have decreased sensitivity to touch unless adventitia is involved

B. Rectum (L1-S4)—stimulated by distention

C. Bladder (S2–S4)—stimulated by distention

Gluteal Region

I. BOUNDARIES—iliac crest, inferior border of the gluteus maximus (at the level of the transverse gluteal fold), lateral aspect of the sacrum and coccyx, and tensor fascia lata

➢ **II. SCIATIC FORAMEN** (330, 331)

A. Greater Sciatic Foramen—located between the greater sciatic notch and the sacrospinous and sacrotuberous ligaments
Transmits:

1. *Piriformis muscle (The sciatic nerve and the posterior cutaneous nerve of the thigh enter the greater sciatic foramen below the piriformis muscle.)*

2. *Superior gluteal vessels and nerves—enter the gluteal region above the piriformis muscle*

3. *The internal pudendal vessels and nerves enter the gluteal region below the piriformis, cross the ischial spine, and reenter the pelvis via the lesser sciatic foramen.*

B. Lesser Sciatic Foramen—Transmits:

1. *The tendon of the obturator internus*

2. *The reentering internal pudendal vessels and nerves*

➢ **III. MUSCLES** (461)

A. Extensor of the Thigh—Gluteus maximus—innervated by the inferior gluteal nerve (L5-S2)

1. *From the posterior pelvis over the trochanteric bursa to the iliotibial tract of the fascia lata*

B. Abductors of the Thigh—innervated by the superior gluteal nerve

1. *Gluteus medius and minimus—from the ilium to the greater trochanter*

2. *Tensor fasciae latae—from the anterior superior iliac spine to the iliotibial tract*

C. External Rotators of the Thigh—innervation via the sacral plexus

1. *Piriformis—from the sacrum and sacrotuberous ligament via the greater sciatic foramen to the greater trochanter*

2. *Obturator internus—from the obturator membrane to the greater trochanter*

3. *Gemelli—from the ischium (ischial tuberosity) to the greater trochanter*

4. *Quadratus femoris—from the ischium to the quadrate tubercle of the femur*

IV. ARTERIES IN THE GLUTEAL AND SACRAL AREAS (465, 468)—Because branches of the aorta, vertebrals, and internal and external iliacs have anastomoses, bleeding due to injury to the gluteal or sacral arteries may be difficult to control.

A. Superior gluteal (247, 320, 468)—from the internal iliac—enters the gluteal region thru the greater sciatic foramen. It may arise from a common trunk along with the inferior gluteal. Possible **anastomoses** with B, F, and G

B. Inferior gluteal (247, 373, 468)—from the internal iliac. It enters the gluteal region thru the greater sciatic foramen (along with the nerve). It lies close to the sacrospinous ligament and the sacrotuberous ligament. Possible **anastomoses** with A, B, C, D, F, G, and H

C. Pudendal (376) from internal iliac—possible **anastomoses** with B and D

D. Branches of the vertebral—possible **anastomoses** with B and C

E. Middle sacral—from the aorta—possible **anastomoses** with F and G

F. Lateral sacral (247)—from the internal iliac—possible **anastomoses** with A, B, E, and G

G. Femoral circumflex arteries (477)—from the profunda femoris (deep femoral)—possible **anastomoses** with A, B, and H

H. Perforating branches of the profunda femoris (deep femoral) (468)—possible **anastomoses** with B and G

V. VEINS (292)—superior and inferior gluteals to internal iliacs

VI. LYMPHATICS—gluteals to internal iliac nodes

VII. NERVES (465, 469)—from the sciatic nerve (L4-S3). The sciatic and lateral femoral cutaneous nerves pass thru the gluteal region.

A. Sensory

1. *Posterior cutaneous nerve of the thigh (S1–S3)—supplies the posterior thigh and lower medial aspect of the buttock*

2. *Cutaneous branches of lumbar and spinal nerves—supply upper and medial aspects of the gluteal region*

B. Motor

1. *Superior gluteal (L4-S1)—to the gluteus medius and minimus and the tensor fascia lata*

2. *Inferior gluteal (L5-S2)—to the gluteus maximus*

3. *Sacral plexus (L4-S4)—to the external thigh rotators*

Bladder

I. SURFACES (342)

A. Superior—related to the ileum and colon and to the uterus in the female

B. Inferolateral—The retropubic space (of Retzius) which contains a prominent venous plexus

C. Posterior (the bladder base)—adjacent to the rectovaginal septum

II. OPENINGS (343)

A. Urachus (Median Umbilical Ligament) (only rarely open)—a fibrous cord connecting the apex of the bladder and the umbilicus

B. Ureters—openings located at the posterolateral aspect of the bladder—internally connected by the interureteric ridge

C. Neck—gives rise to the urethra

1. *Connected to the pubis by the pubovesical (female) or puboprostatic (male) ligaments*
2. *Male—related to the prostate*
3. *Female—The bladder neck is related to the upper part of the UG diaphragm.*

III. LAYERS (343)

A. Serous—on the upper surface of the bladder

B. Fascia—Endopelvic fascia around the bladder aids in support.

C. Muscular

1. *Smooth muscle which extends in different directions and extends to the neck to form the sphincter and then to the urethra (female) or prostate (male)*
2. *Active contractions of the detrusor (smooth) muscle decrease the length of the urethra and increase the caliber of the urethral lumen.*

D. Submucous—connective tissue between the muscularis and mucosa

E. Mucosa—Transitional epithelium forms folds when the bladder is empty, except for the trigone (a triangular area on the posterior wall formed by the ureteral and urethral orifices).

IV. ARTERIES (319, 320)

A. Superior Vesical (the unobliterated part of the umbilical artery)—from the internal iliac

B. Middle Rectal—supplies the posterior surface of the bladder

C. Inferior Vesical—supplies the inferior bladder and prostate (male)—from the internal iliac

D. Vaginal (Female)—from the internal iliac

V. VEINS (248)—to the vesical plexus, then to the inferior vesical veins

VI. LYMPHATICS (321)

A. Anterior portion of the bladder—mainly to external iliac nodes

B. Posterior portion of the bladder—mainly to internal iliac nodes

VII. NERVES (322, 388)

A. Sympathetic (L1–L2)—via vesical plexus, pelvic plexus, and hypogastric plexus

Stimulation of fibers from the hypogastric nerve closes the urethrovesical junction, allows seminal emission, and prevents retrograde ejaculation.

Sympathetic mediation may depend not on direct effects of noradrenergic fibers on the detrusor but on indirect inhibition of excitatory parasympathetics within the ganglia of the pelvic plexus.

B. Parasympathetic (S2–S4)—to the detrusor muscle via the pelvic nerve

Loss of parasympathetic innervation → atonic bladder

C. Visceral Afferents—along sympathetic and parasympathetic pathways

D. Somatic—(motor) via the pudendal nerve (S2–S3) to the external sphincter

Urethra—Male

I. DIVISIONS (359)

A. Prostatic—above the superior fascia of the UG diaphragm

1. *Widest and most dilatable—length about 1 inch*
2. *A derivative of the urogenital sinus*
3. *Openings*

   i. ejaculatory ducts—open into the seminal colliculus

   ii. prostatic ductules—open into the prostatic sinuses located on the sides of the crest (a ridge on the posterior urethral wall). The highest point of the crest is called the seminal colliculus.

   iii. prostatic utricle

      a. a small diverticulum formed by the fused ends of the paramesonephric ducts

      b. opens into the prostatic urethra on the colliculus

      c. homologous to the uterus and vagina ("the male vagina and uterus")

B. Membranous Urethra

1. *Length just under 1 inch—the shortest part of the urethra*
2. *A derivative of the urogenital sinus*
3. *Lies within the UG diaphragm*
4. *Surrounded by the sphincter urethrae (external urethral sphincter). Voluntary fibers come from the UG diaphragm.*
5. *Least dilatable and narrowest part*

C. Spongy Urethra (Penile Urethra)—the portion of the urethra beyond the perineal membrane

1. *Length, 6 inches; diameter, 5 millimeters*
2. *Surrounded by erectile tissue*
3. *The end of the spongy urethra dilates into the fossa terminalis (navicular fossa).*

The most common site of urethral rupture is the area below the perineal membrane (superficial covering of the UG diaphragm). Urine can go into the superficial perineal space (superficial pouch) → scrotum.

II. MUSCLES (359)

A. Smooth muscle around the upper portions

B. Urethral Sphincter—surrounds the membranous urethra

1. *From the levator and UG diaphragm*
2. *Contraction of the voluntary sphincter around the membranous urethra can stop the urinary stream.*
3. *Bulbospongiosus—compresses urethra and contributes to penile erection*

III. ARTERIES (357, 374, 376)

A. Proximal Urethra—prostatic artery from the inferior vesical

B. Distal Urethra—urethral artery from the artery of the penis from the pudendal

IV. VEINS—corresponding to arteries

V. LYMPHATICS (379)

A. Prostatic Urethra—to the internal iliac nodes

B. Membranous Urethra—to the internal iliac nodes

C. Penile Urethra—Lymphatics accompany those of the glans penis along the dorsal surface of the penis to the external iliac nodes.

VI. NERVES

A. Somatic—Pudendal nerve. Somatomotor innervation of the sphincter thought to be via the pudendal nerve may be partially innervated autonomically.

B. ANS—via the prostatic plexus via the inferior hypogastric plexus

1. *Sympathetic—contracts the sphincter*
2. *Parasympathetic—relaxes the sphincter*

C. Visceral Afferents—along with pelvic splanchnics

VII. GLANDULAR OPENINGS INTO THE URETHRA (359)

A. Urethral Mucous glands

B. Ducts of the Bulbourethral glands

C. Prostate

D. Seminal Vesicles

Urethra—Female

The proximal ¾ of the female urethra is called the neck. The epithelium is derived from endoderm. The upper portion of the female urethra is lined with transitional epithelium.

I. MUSCLES (343, 353)

A. Voluntary Muscles—sphincter of the urethra—including fibers from the levator and deep transverse perineal muscles

The angle (urethrovesical) between the bladder and urethra is important in maintaining urinary continence. An increase in the angle may result from neuromuscular damage (local partial denervation) of the sphincters during childbirth and aging. Restoration of the angle may be attempted by sphincter strengthening ("Kegal" type exercises) and/or surgery.

B. Smooth Muscle—**and connective tissue around the proximal urethra helps to prevent leakage of urine**

II. GLANDS (353)

A. Urethral Glands—mucous

B. Paraurethral Glands (Skene's)—Ducts open alongside the external urethral orifice. (Skene's glands are homologous to the male prostate.)

III. ARTERIES

A. Urethral—from the artery to the clitoris from the internal pudendal

B. Vaginal—from the anterior division of the internal iliac

IV. VEINS—correspond to arteries

V. LYMPHATICS—to the internal iliac and sacral nodes

VI. NERVES (384)

A. Somatic—Pudendal nerve. Perineal branches of the pudendal (S2–S4) innervate the external sphincter.

B. Sympathetic (L1–L2)—contract the sphincter and contribute to urinary retention

C. Parasympathetic (S2–S4)—relax the involuntary sphincter

D. Visceral Afferents—along with pelvic splanchnics

Micturition

Micturition involves the storage and periodic evacuation of urine.

I. RECEPTORS

A. Fundus—Beta adrenergic receptors are concentrated in the fundus (tend to relax the bladder).

B. Trigone and Urethra—Alpha adrenergic receptors are concentrated in the trigone and urethra.

II. MUSCULAR ACTION

A. Smooth Muscle and connective tissue are necessary for continence. The proximal 3 centimeters of urethra prevents leakage due to tension of smooth muscle and connective tissue.

B. Bladder fills, then stretches, then contracts, then stretches, then contracts.

C. There is Enhanced Efficiency—with the aid of voluntary (striated) muscle around the urethra from the levator ani and urogenital diaphragm—contraction can stop the stream—innervation carried via the pudendal nerve. Voluntary contractions of the levator ani can increase urethral length up to 5 millimeters.

D. The External Urethral Sphincter consists of type I (slow twitch) fibers for tonic contraction

III. PRESSURE—constant until the bladder is full (300+ cc), then a sharp rise, then the sensation of necessity to void. If intravesical pressure exceeds the ability of the sphincter to hold, loss of urine occurs.

IV. DECREASED SPHINCTER RESISTANCE—Is due to:

A. Relaxation of striated muscle—decreases the length of the urethra

B. Active contraction of the bladder—decreases the length of the urethra and increases the caliber of the lumen

V. ABNORMALITIES OF MICTURITION

A. Atonic Bladder—loss of muscle tone due to chronic overstretching

B. Detrusor Areflexia—decreases motor innervation

C. Detrusor Hyperreflexia—loss of higher center control

D. Unstable Bladder—involuntary contraction

Male Reproductive System

I. TESTES (338, 361, 362)

A. Coverings (362)

1. *Tunica vaginalis—located anteriorly and on the sides. It is a closed sac embryologically derived and cut off from the peritoneum. The tunica covers the testis and epididymis.*

   A hydrocele is fluid collection in the tunica vaginalis.

2. *Tunica albuginea—connective tissue coat over the entire testis. Posteriorly, it extends inward to divide the testis into lobes.*

B. Arteries (314, 372)—testicular from the aorta

C. Veins (372)

1. *Right testicle—pampiniform plexus → testicular → IVC*
2. *Left testicle—pampiniform plexus → testicular → left renal*

   Pampiniform plexus dilation is called a varicocele.

D. Lymphatics (379)—to lumbar and aortic nodes

E. Nerves (387)

1. *Sympathetic (SNS)—from renal and aortic plexus*
2. *Visceral afferents accompany Sympathetic fibers*

   Pain can be referred to the upper abdomen (e.g., testicular trauma → nausea).

II. EPIDIDYMIS (362)—for the storage and maturation of sperm

A. Location—posterosuperior aspect of the testis

B. Divisions

1. *Head—fits over upper part of the testis*
2. *Body—separated from the testis by the epididymal sinus*
3. *Tail—stores spermatozoa—gives rise to the vas*

C. Arteries—testicular from the aorta

D. Veins—pampiniform plexus → testicular veins

E. Lymphatics—to lumbar nodes

F. Nerves—same as for the testes

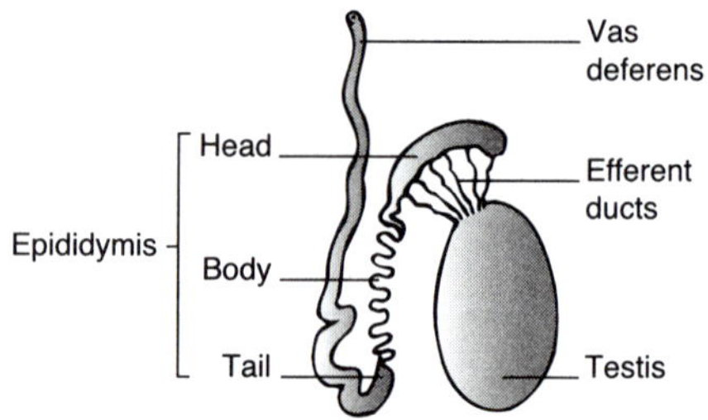

III. VAS DEFERENS (338, 362)—a thick muscular tube

A. Course—the vas begins at the lower end of the epididymis → over the back of the testis into the inguinal canal as part of the spermatic cord → leaves the deep inguinal ring → over external iliac vessels → along the lateral wall of the pelvis → posterior to

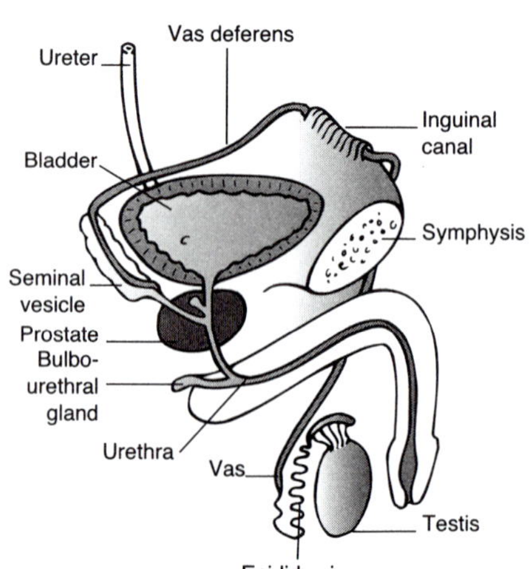

bladder → medial to the seminal vesicle → dilates to form the ampulla (for sperm storage) → joins the duct of seminal vesical to form the ejaculatory duct

B. Arteries—superior vesical

C. Veins—correspond to arteries

D. Lymphatics—to the external iliac nodes

E. Nerves—inferior hypogastric plexus

IV. SEMINAL VESICLES (358)—derived from the mesonephric duct—provide alkaline seminal fluid with a high fructose content. Each vesicle is 5–6 centimeters long but is 15 centimeters long when its coils and convolutions are unraveled.

A. Location—rest in fat on the levator ani muscle—between bladder and rectum. They are partially covered by peritoneum.

B. Layers

1. *Low columnar mucosal epithelium and lamina propria connective tissue*
2. *Muscularis—2 layers—contracts during sex*
3. *Adventitia*

C. Arteries—inferior vesical and middle rectal

D. Veins—corresponding

E. Lymphatics—to external and internal iliac nodes

F. Nerves

1. *Sympathetic—from the superior lumbar nerves*
2. *Parasympathetic—pelvic splanchnics*

V. EJACULATORY DUCT (338)

A. Formed by the vas and the ducts of the seminal vesicles—extends thru the prostate gland

B. Opens into the seminal colliculus of the prostatic urethra

VI. PROSTATE (358)—A walnut-sized organ which supplies ⅕ of the seminal fluid. It surrounds the urethra and is located between the bladder and rectum.

A. Anterior Fibromuscular Stroma—constitutes ⅓ of the prostate. It is composed of collagen, elastin, and muscle, and it connects with the prostatic capsule.

B. Capsule

C. Zones—divide the glandular elements according to their pathology and the locations of their ducts

The zonal classification is used by urologists.

1. *Transition zone—surrounds the urethra proximal to the ejaculatory ducts. The transition zone comprises 5–10% of the glandular prostate.*
   i. The ducts of this zone arise beneath the preprostatic sphincter then pass beneath it and extend along its lateral and posterior sides.
   ii. It is surrounded by fibromuscular tissue which separates it from the other glandular portions.
2. *Central zone—surrounds the ejaculatory ducts and extends under the base of the bladder. The central zone accounts for 25% of the glandular prostate.*
   i. The ducts of this zone arise around the openings of the ejaculatory ducts.
   ii. The central zone accounts for 25% of the glandular prostate.
3. *Peripheral zone—covers most of the apical, posterior, and lateral aspects of the prostate. The peripheral zone accounts for 70% of the glandular prostate.*
   i. Its ducts drain into the prostatic sinuses.
   ii. The peripheral zone accounts for 70% of the glandular prostate.
   iii. The zone most likely to become pathologic

D. Divisions (Lobes)—Some anatomy texts divide the prostate into 2 lateral lobes and 1 middle lobe while others include an anterior lobe. The lobe classification is used by anatomists.

1. *Anterior lobe—anterior to the urethra*
2. *Lateral lobes—are lateral to the urethra, form the base of the prostate, and are separated by a central sulcus*
3. *Middle (Median) lobe—The glandular lobe is located between the urethra and ejaculatory ducts.*
   i. prostatic utricle—blind diverticulum
   ii. seminal colliculus—opening for the ejaculatory ducts
4. *Posterior lobe—posterior to the urethra*

E. Coverings

1. *Fibrous capsule*
2. *A visceral layer of pelvic fascia is called the* ***Prostatic Sheath****. Posteriorly, it joins with the rectovesical septum.*

F. Arteries—inferior vesical and middle rectal from internal iliac

1. *Inferior vesical—from the internal iliac*

i. urethral—to periurethral glands and transition zones

ii. capsular—courses posterolaterally

2. *Middle rectal*

G. Veins—The prostatic venous plexus (between the fibrous capsule and prostatic sheath) extends below the ureter to terminate in the internal iliac vein and into the vesical and vertebral plexuses.

➢ H. Lymphatics (379)—to internal iliac, obturator, and sacral nodes. Some extend over the bladder to go to the external iliacs.

➢ I. Nerves (388)—in the hypogastric plexus

1. *Sympathetics—via inferior hypogastrics—contract smooth muscle*

2. *Parasympathetics—via pelvic splanchnics—increase secretion*

3. *Visceral afferents—via pelvic plexuses to the spinal cord*

➢ VII. PROSTATIC URETHRA (359)

A. Urethral Crest—a ridge on the posterior wall of the urethra. The highest point of the crest is called the seminal colliculus.

B. Prostatic Utricle—a cul-de-sac which opens into the colliculus

C. Prostatic Sinus—a groove on each side of the crest which receives prostatic ducts

➢ VIII. SEMINAL FLUID SECRETIONS (338)

A. Seminal vesicles—fluid contains fructose and potassium

B. Prostate—alkaline, milky-like, serous fluid containing coagulation enzymes, sperm motility enhancers, and acid phosphatase

C. Bulbourethral (Cowper's)—located in the UG diaphragm. Secretions lubricate the urethra prior to ejaculation.

D. Urethral—mucus secreting

➢ IX. COURSE OF SPERM (338)—seminiferous tubules → rete testes → efferent ductules → head of epididymis → tail of epididymis → vas deferens → ejaculatory duct → prostatic urethra → membranous urethra → penile urethra

Female Reproductive System

I. VAGINA (337)

A. Vaginal Fornices (337)—formed by the cervix

1. *Anterior fornix—shallow*
2. *Posterior fornix—larger*
3. *Two lateral fornices*

B. Supports (333, 334, 337)

1. *Upper ⅓—transverse cervical (cardinal) ligaments*
2. *Middle ⅓—levator ani*
3. *Lower ⅓—levator ani*

C. Relationships (337, 344)

1. *Upper anterior—bladder and ureters*
2. *Upper posterior—very close to posterior cul-de-sac* ***(important for drainage and ultrasound)***
3. *Middle posterior—rectum*

D. Arteries (373, 375)

1. *Upper ⅓—vaginal (from the uterine or internal iliac)*
2. *Middle ⅓—internal pudendal*
3. *Lower ⅓—middle and inferior rectal*

E. Veins—to the internal iliacs

F. Lymphatics

1. *Upper ⅓—to internal iliac nodes and sacrals*
2. *Middle ⅓—to internal iliac nodes*
3. *Lower ⅓—to vulvar and superficial inguinal nodes*

II. UTERUS (337, 345–348)

A. Dimensions—9 centimeters long, 6 centimeters wide, 4 centimeters anteroposteriorly

B. Weight—nulliparous, 40–50 grams; multiparous, 70–80 grams

C. Divisions

1. *Body—upper ⅔ of the uterus*
    - i. fundus—the area above the tubes
    - ii. isthmus—the area between body and cervix

2. *Cervix—the lower ⅓ of the uterus. (The cervix contains a **transition (transformation) zone** with columnar epithelium above and squamous epithelium below.)*

D. Endometrium

1. *Functional zone—sheds during menstruation*

   i. Stratum compactum—superficial

   ii. Stratum spongiosum—intermediate

2. *Stratum basalis—remains during menstruation*

➢ E. Relationships (344, 348)

1. *Anterior—bladder*
2. *Posterior—rectum*
3. *Inferior—vagina*
4. *Lateral—broad ligaments and ureters. The ureter passes underneath the uterine artery and less than 1 inch from the cervix.*

F. Arteries (The uterine and ovarian arteries have **anastomoses.**)

1. *Uterine—from the internal iliacs*

   i. cervical

   ii. vaginal—the descending branch

2. *Ovarian—from the aorta—supplies the uterine fundus*

G. Veins—uterine to internal iliacs

➢ H. Lymphatics (377)

1. *Fundus—to aortic, lumbar, and iliac nodes*
2. *Remainder of body—to iliac nodes*
3. *Cervix—to sacral, rectal, and iliac nodes*

➢ I. Nerves (383, 385, 386)

1. *Uterine body*

   i. sympathetic—inferior hypogastric plexus

   ii. parasympathetic—pelvic splanchnics—S2–S4

   iii. visceral afferents—in the uterosacral ligaments and thence via inferior hypogastric plexus to T10-L1

2. *Cervix—visceral afferents to S2–S4*
__NOTE:__ Ganglia located behind and lateral to the cervix are sometimes call Frankenhäuser's ganglia.

III. TUBES (OVIDUCTS) (346)

A. Divisions and Dimensions (346)

1. *Intramural (Interstitial)—the portion of the tube within the wall of the uterus—2 centimeters long*
2. *Isthmus—4 centimeters (inside diameter, 1–2 millimeters)*
3. *Ampulla—the thin-walled portion—5 centimeters long (inside diameter, 6 millimeters)* ***(the location of most tubal pregnancies)***
4. *Infundibulum (funnel-shaped distal portion)—has* **Fimbriae** *(finger-like projections)*

B. Arteries (375)—ovarian from the aorta

C. Veins—ovarian and uterine

D. Lymphatics—Some go via ovarian lymphatics to the aortic nodes and others go via uterine lymphatics to the internal iliacs.

E. Nerves (383, 386)

1. *Sympathetic—ovarian plexus from the aortic, renal, and hypogastric plexuses*
2. *Parasympathetic—from vagal fibers in the ovarian plexus*
3. *Visceral afferents to T11-L1*

Tubal expansion due to infection or tubal pregnancy may initially produce lower abdominal pain.

IV. OVARIES (346, 349)

A. Dimensions, 3.5 × 2.5 × 1.5 centimeters; weight, 4–9 grams

B. Divisions (349)

1. *Germinal epithelium—a single cuboidal layer*
2. *Cortex—outer zone—contains follicles and is covered by a single cuboidal layer of germinal epithelium*
   - i. tunica albuginea—connective tissue beneath the surface (germinal) epithelium
   - ii. stroma—connective tissue and follicles
3. *Medulla—inner zone—contains connective tissue surrounding blood vessels*
4. *Hilum—located at the insertion of the mesovarium—the entrance for blood vessels and nerves*

C. Supports

1. *Mesovarium—part of the broad ligament*

2. *Ovarian ligament—from the ovary to the uterus*

D. Arteries (371, 375)—ovarian from the aorta

E. Veins—The right drains into the IVC. The left drains into renal.

F. Lymphatics (377)—to lumbar, aortic, and internal iliac nodes

G. Nerves (386)

1. *Sympathetic—T5–T12*

2. *Visceral afferents to T10–T11 or higher*

V. SUPPORTS OF THE FEMALE REPRODUCTIVE ORGANS (344–346)

A. Peritoneal Supports (346)—consist of the subdivisions of the broad ligament. The broad ligament consists of 2 layers of pelvic peritoneum extending from the uterus to the lateral pelvic wall.

1. *Mesosalpinx—the reflection of broad ligament over the tube.*

2. *Mesovarium—extends from the anterior portion of the ovary to the posterior leaf of the broad ligament*

3. *Suspensory ligament of the ovary (Infundibulopelvic ligament)—contains the ovarian vessels and nerves*

4. *Utero-ovarian—from the inferior part of the ovary to the uterus*

B. Connective Tissue Supports (341, 346)

1. *Transverse cervical (Cardinal) ligament—extends from the cervix to the lateral pelvic wall*

2. *Uterosacral ligament—extends from the cervix posterolaterally to the sacrum*

3. *Parametrium—connective tissue alongside the uterus. The upper parametrial tissue contains the tube, ovarian, and round ligament. The lower portion contains the uterine vessels. Anteriorly and posteriorly, the parametrium is covered by the peritoneum of the broad ligament, which in turn extends laterally to the lateral wall of the pelvis as the mesometrium of the broad ligament.*

C. Smooth Muscle Support (344, 345)

1. *Round ligament—extends from the superior side of the uterus thru the inguinal canal to the labium majus. Its attachment to the uterus is just inferior and anterior to the attachment of the oviduct into the uterus.*

Perineum—General

I. BOUNDARIES (354) (The perineum is an area below the pelvic diaphragm.)

A. Superior—muscles of the pelvic diaphragm

B. Anterior—ischial rami, inferior pubic rami, and symphysis pubis

C. Posterior—ischial tuberosities, sacrotuberous ligaments, and coccyx

II. DIVISIONS (354)—The perineal area consists of two triangles divided by a line between the ischial tuberosities.

A. Urogenital Triangle—contains external genitalia and terminal parts of urogenital passages

B. Anal Triangle—contains the anal canal with ischioanal (ischiorectal) fossa on each side

III. SUPPLY (364, 382)—The perineum is generally supplied by the pudendal vessels and nerves which travel to the perineum in the pudendal canal. (The testicles are supplied by testicular vessels and nerves.)

IV. PELVIC DIAPHRAGM (333–337)—the superior limit of the perineum

A. Separates the pelvic viscera from the ischiorectal fossa

B. Consists primarily of the levator muscles, the major support of the pelvic viscera

V. UROGENITAL DIAPHRAGM (352, 357)—consists of muscles and their fascial coverings

A. Stretches between the two sides of the pubic arch

1. *Some fibers extend between the ischial tuberosities as the deep transverse perineal muscles.*
2. *Other fibers extend around the membranous urethra as its external (voluntary) sphincter.*

B. Has two fascial coverings (layers or sheets)

1. *Superficial (Inferior)—also called the* ***perineal membrane.*** *It is absent immediately under the pubic symphysis.*
2. *Deep (Superior)—membranous—a continuation of Scarpa's fascia*

Perineum—Male

The perineum (354–356) consists of urogenital and anal regions.

I. UROGENITAL REGION (338)—The urogenital region includes the membranous urethra, scrotum, and penis.

A. Penis (355, 356)

Subcutaneous dorsal vein
Deep dorsal vein
Corpus cavernosum
Dorsal artery
Buck's fascia
Superficial fascia
Corpus sponglosum
Penile urethra

1. *Major Parts (3)*
   - i. Corpus spongiosum—contains the penile urethra
     - a. proximal—called the "bulb"
     - b. distal—glans
   - ii. Corpora cavernosa—paired—attached to the pubic arch—covered by the ischiocavernosus muscles
     - a. proximal—crura
     - b. distal—pendant
2. *Other parts*
   - i. prepuce—a continuation of penile skin
   - ii. frenulum—portion of skin attached to the prepuce
3. *Fascia*
   - i. superficial subcutaneous—contains superficial veins and lymphatics—continuous with the tunica dartos (dartos fascia)
   - ii. deep (Buck's fascia)—contains the deep dorsal vein, nerves, and lymphatics
4. *Muscles*
   - i. bulbospongiosus (bulbocavernosus) muscles
     - a. extend from the central tendon (perineal body) to the dorsum (deep fascia) of the penis
     - b. contract at termination of micturition
   - ii. ischiocavernosus muscles—from the ischial tuberosities to envelop the crura of the penis and aid in maintaining penile erection
   - iii. superficial transverse perineal muscles—from the ischial tuberosities to the bulb of the penis to the central tendon

5. *Arteries (376)—perineal branches of the pudendal*
   - i. dorsal artery of the penis—between the corpora cavernosa
   - ii. deep arteries of the penis—within the corpora cavernosa

6. *Veins* (336, 355, 374, 376)
   i. deep dorsal vein—to the prostatic plexus
   ii. superficial dorsal vein—to superficial external pudendal veins
7. *Lymphatics—from superficial perineal areas to the inguinal nodes—from deep perineal areas to the internal iliac nodes and from the testes to para-aortic nodes*
8. *Nerves* (376)*—the dorsal nerve and the perineal nerve from the pudendal nerve*

B. Scrotum (361, 374)

1. *Layers*
   i. superficial fascia—a continuation of Camper's fascia—2 layers of superficial fascia cover the dartos muscle
   ii. external spermatic fascia—a continuation of the external oblique muscle
   iii. cremasteric fascia—covers the cremaster muscle
   iv. internal spermatic fascia—a continuation of the transversalis fascia
   v. tunica vaginalis—covers the front and sides of the testis. It is a closed sac. The sac is a remnant of the processus vaginalis and therefore has parietal and visceral layers.
2. *Muscles—attached to the scrotal skin*
   i. cremaster—a continuation of the internal oblique muscle
   ii. dartos—covered by fascia which is continuous with the superficial perineal fascia and the subcutaneous tissue of the anterior abdominal wall
   NOTE: The genital branch (motor) of the genitofemoral nerve (L1–L2) innervates the cremaster muscle.

   The femoral branch (sensory) of the genitofemoral nerve innervates the anteromedial portion of the thigh.

   This innervation forms the basis of the "cremasteric reflex."
3. *Arteries* (376)
   i. perineal—from the pudendal from the internal iliac
   ii. superficial external pudendal (from the femoral)—supplies the scrotum
4. *Veins—scrotal veins drain into the pudendal veins*
5. *Lymphatics—to superficial inguinal nodes*

6. *Nerves (380, 382)*
   i. anterior scrotum—from the ilioinguinal (L1)
   ii. anteromedial scrotum—from the femoral branch of the genitofemoral (L1–L2)
   iii. posterior scrotum
      a. scrotal branches of the perineal nerve of the pudendal nerve
      b. perineal branch of posterior femoral cutaneous nerve

C. Urogenital Diaphragm (358)—2 layers of fascia and 2 muscles—The area within the UG diaphragm is sometimes called the **deep perineal pouch (space)**.

1. *Muscle fibers between the sides of the pubic arch*
   i. sphincter urethrae—voluntary sphincter around membranous urethra
   ii. deep transverse perineal muscle—fibers extend to the central tendon
2. *Fascia*
   i. superior (deep) (upper) layer (a continuation of Scarpa's fascia)
   ii. inferior (superficial) (lower) layer **(perineal membrane)** (a continuation of Camper's fascia). Passing thru the inferior fascial layers are:
      a. urethra
      b. ducts of bulbourethral glands
      c. arteries to the corpora cavernosa
      d. internal pudendal artery—the terminal branch of the internal iliac
      e. dorsal nerve of the penis

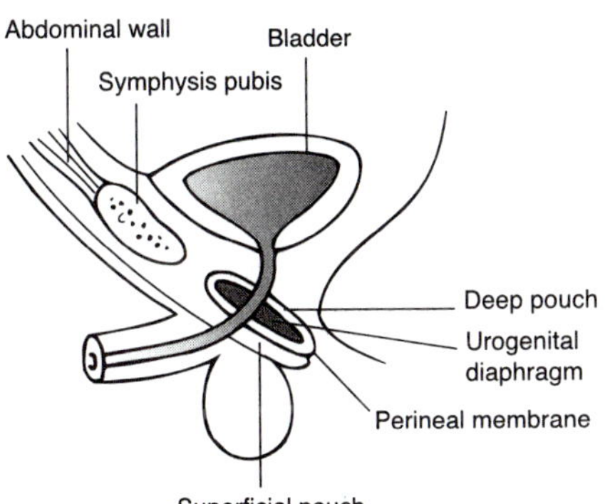

3. *Innervated by the perineal nerve—from the pudendal*
4. *Bulbourethral (Cowper's) glands (358)—located in the UG diaphragm, but their ducts extend thru the inferior fascial layer (perineal membrane) to enter the urethra. They produce ejaculatory secretions.*

D. The Area Below the Urogenital Diaphragm (357) is sometimes called the **superficial perineal pouch.** The pouch contains:

1. *The root of the penis—corpora cavernosa*

Chapter 6 *Pelvis*

2. *Muscles covering the penile part of the urethra* (355)

   i. bulbocavernosus—contracts at the end of micturition

   ii. ischiocavernosus—covers the crus of the penis and results in erection and erectile maintenance

3. *Superficial transverse perineal muscle* (356) *extends from the ischial tuberosities to the perineal body (central tendon of the perineum)*

4. *Perineal branch of pudendal nerve—supplies muscles and skin*

E. Male Central Tendon (Perineal Center) (356)—is formed by:

1. *External anal sphincter*
2. *Bulbospongiosus (Bulbocavernosus) muscle*
3. *Superficial transverse perineal muscle*
4. *Posterior fibers of the deep transverse perineal*
5. *Levator fibers which support the prostate gland*

II. ANAL REGION (354, 364)

A. Anal Triangle (354). Its sides are:

1. *A line between the ischial tuberosities*
2. *Lines between the tip of the coccyx and the ischial tuberosities*

B. Anal Canal—located inferior to the pelvic diaphragm

C. Ischioanal (Ischiorectal) Fossa (364)

1. *Bordered by the levator ani muscle, obturator internus muscle, ischial tuberosity, external anal sphincter, sacrotuberous ligament, and lesser sciatic foramen*

2. *Contents*

   i. pudendal canal (Alcock's) (364)—located in the lateral wall of the ischioanal fossa and covered by the obturator internus fasci—contains the pudendal vessels and nerves

   ii. fat—allows dilation of the canal during defecation

   iii. inferior rectal (hemorrhoidal) vessels and nerves

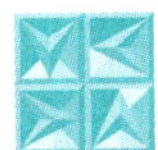

Perineum Female

I. UROGENITAL REGION (351, 352)

A. Vulva (350)—superficial tissues of the urogenital region including the mons, labia, and clitoris

1. *Labia majora—(homologous to the scrotum of the male)*
   - i. skin overlying fat
   - ii. the junction of the labia majora above the clitoris (over the symphysis) is called the anterior commissure
   - iii. the junction of the labia majora posteriorly is called the posterior commissure
   - iv. contain hair follicles, sebaceous glands, and sweat glands
2. *Labia minora—skin between the labia majora and the vaginal orifice*
   - i. enclose the **vestibule** of the vagina which contains the vaginal, urethral, and greater vestibular gland openings
   - ii. meet anteriorly at the clitoris and posteriorly in the inner aspects of the labia majora

     Clinically, the anterior meeting of the labia minora is called the prepuce and the posterior meeting of the labia minora with the labia majora is called the fourchette.
3. *Hymen—thin area covering the entrance to the vagina*
   - i. imperforate hymen—does not have an opening
   - ii. cribriform hymen—has small perforations
   - iii. After it is stretched during childbirth, it forms tags called **carunculae myrtiformes**
4. *Clitoris—(homologous to the penis)—consists of a glans, body (corpus), and corpora cavernosa. The clitoris does not contain a corpus spongiosum.*
   - i. body (corpus)—formed by the junction of the corpora cavernosa
   - ii. The corpora cavernosa (erectile tissue) continue laterally from the root of the clitoris to either side of the vagina where they are known as the vestibular "bulbs" (analogous to the corpus spongiosum of the penis).
   - iii. The clitoral crura are continuations of the corpora cavernosa which diverge from the body and extend along the ischial (pubic) rami.
   - iv. The crura and corpora join inferior to the pubic symphysis
   - v. Parasympathetic nerves are involved in clitoril erection.

Chapter 6 *Pelvis*

5. *Urethral opening—below it are openings to the paraurethral (Skene's) ducts which are homologous to the male prostate*
6. *Vaginal opening*
   i. superficial to the hymen
   ii. has openings for mucus-secreting glands called **the greater vestibular glands (Bartholin's)**
      a. analogous to Cowper's glands in the male
      b. pea-sized glands with ducts about 1 centimeter long
      c. The duct openings are located between the hymen and labia minora at 4 and 8 o'clock.
7. *Mons—a pad of fat over the symphysis*

➢ B. Muscles in External Urogenital Region (352)

1. *Bulbospongiosus (bulbocavernosus)—From the central tendon (perineal body) they extend on each side of the vaginal opening (covering the corpora cavernosa) to insert on the clitoris.*
2. *Ischiocavernosus—from the medial portion of the ischial rami to insert on (and therefore compress) the crura*
3. *Superficial transverse perineal—ischial tuberosities to the central tendon*

➢ C. Urogenital Diaphragm (UG Diaphragm) (352)

1. *Surrounds urethra, vagina, and rectum*
2. *Stretches between the two sides of the pubic arch*
3. *Has two fascial layers which enclose the muscles*
   i. deep layer—membranous—a continuation of Scarpa's fascia
   ii. superficial layer—also called the **perineal membrane**—a part of Colles' fascia and a continuation of Camper's fascia
   iii. Anteriorly and posteriorly, the layers fuse.
4. *The space between the superficial and deep fascial layers is sometimes called the **deep perineal pouch**. It contains:*
   i. part of the urethra and vagina
   ii. the deep transverse perineal muscle and urethral sphincter. (The urethral sphincter is sometimes listed as a separate muscle but still part of the UG diaphragm.)
   iii. internal pudendal vessels
   iv. dorsal nerve of the clitoris

D. The Area Below the UG Diaphragm (351)—occasionally described as the **superficial perineal pouch**. This portion of the perineum contains:

1. *Clitoril crura*
2. *Bulbospongiosus and ischiocavernosus muscles*
3. *Superficial transverse perineal muscle—from the ischial tuberosities to the perineal body*
4. *Perineal branch of the pudendal nerve which supplies the above muscles and skin*
5. *Perineal body (Central tendon)*
6. *Greater vestibular gland*

E. Internal Pudendal Artery (373, 375)—supplies the female perineum

1. *Inferior rectal—to the anal canal*
2. *Perineal—to the posterior labia and the superficial perineal pouch*
    i. posterior labial
    ii. branch to ischiocavernosus muscle
    iii. branch to bulbospongiosus (bulbocavernosus) muscle
3. *Artery to the clitoris—in the deep perineal pouch*
    i. deep artery to the clitoris (artery to the bulb)
    ii. dorsal artery to the clitoris

F. Internal Pudendal Veins (364, 370)—drain into the internal iliac

II. ANAL REGION (364)

A. Anal Triangle—formed by:

1. *A line between the ischial tuberosities*
2. *Lines between the tip of the coccyx and the ischial tuberosities*

B. Anal Canal—inferior to pelvic diaphragm

➢ C. Ischioanal (Ischiorectal) Fossa (364)

1. *Bordered by the levator ani muscle, obturator internus muscle, ischial tuberosity, external anal sphincter, sacrotuberous ligament, and lesser sciatic foramen*
2. *Contents*
   - i. pudendal canal (Alcock's)
     - a. contains pudendal nerve (S2–S4) and vessels
     - b. covered by obturator internus fascia
   - ii. fat—allows dilation of the anal canal during defecation
   - iii. inferior rectal (hemorrhoidal) vessels and nerves

➢ III. PERINEAL BODY (CENTRAL TENDON, PERINEAL CENTER, CENTRAL TENDINOUS POINT) (351, 375)

A. The Perineal Body is formed by insertions of the following muscles:

1. *Superficial transverse perineal muscle—from the ischial tuberosities*
2. *Posterior fibers of the deep transverse perineal muscle—from the ischial rami*
3. *Bulbospongiosus (Bulbocavernosus) muscles—from the body of the clitoris and thence around the vaginal opening*
4. *External anal sphincter—from the anus*
5. *External urethral (voluntary) sphincters—from the urethra*
6. *Fibers from the puborectal portion of the levator ani*

Clinicians sometimes call the tissue between the vaginal orifice and the anus the perineum.

Rectum

I. LOCATION (366)

A. Begins at the end of the sigmoid colon and anterior to the 3rd sacral vertebra

B. Ends at the beginning of the anus at the level of the levator ani (puborectalis portion) in front of the coccyx. The terminal (widest) portion is called the **ampulla**. Filling creates the urge to defecate.

II. PERITONEAL RELATIONSHIPS (368)

A. Upper ⅓—front and sides covered by peritoneum

B. Middle ⅓—anterior surface covered by peritoneum

C. Lower ⅓—no peritoneal covering

III. FOLDS (365)—3 transverse folds (superior, middle, and inferior) correspond to indentations at the rectal flexure. The folds were formerly called Houston's valves.

IV. MUSCLES (366)

A. Longitudinal Layer—a continuation of the taeniae but does not have true taeniae, sacculations, or appendices epiploicae

B. Circular Layer

V. ARTERIES (369)—have anastomoses with each other

A. Superior Rectal (Hemorrhoidal)—a continuation of the inferior mesenteric

B. Middle Rectal—from the internal iliacs

C. Inferior Rectal—from the pudendal

VI. VEINS (370)—correspond to arteries. The veins have portosystemic venous anastomoses.

A. Superior Rectal—drain into the portal system

B. Inferior Rectal—drain into systemic veins

VII. LYMPHATICS (297)—follow the arteries to the inferior mesenteric, sacral, and internal iliacs and the superficial inguinal nodes

VIII. NERVES (306, 383)

A. Sympathetics and Parasympathetics via the inferior hypogastric plexus

B. Visceral Afferents (Sensory) (L1-S4)—follow parasympathetics and are stimulated by distention

Anus

The anus is that portion of the alimentary canal below the levator ani muscle.

I. PECTINATE (DENTATE) LINE (365)—a wavy comb- (pecten-) shaped line located at the inferior part of the anal columns. The anal area above the line has a different vascular, lymphatic supply and innervation than the area below the line. The middle rectal veins contribute to the portosystemic anastomoses in this area.

A. Above the Line—that portion derived from the hindgut. The epithelium is mixed columnar and squamous.

1. *Features—Some texts describe the area above the line as the terminal portion of the rectum.*

    i. anal columns (Morgagni's columns)—mucosal folds in the upper portion of the anal canal

    ii. anal sinuses (crypts)—between columns and behind valves

    iii. anal glands—secrete mucus and open into the sinuses

    iv. anal valves—transverse epithelial folds at base of and between the columns. The inferior portions of the valves form the pectinate line.

2. *Arteries (287)—the superior rectal artery which is a direct continuation of the inferior mesenteric*

3. *Veins (365, 370)—The superior rectal veins form an internal venous plexus (located between the mucosa and circular muscle layer). The superior rectal veins drain into the inferior mesenteric vein (portal system).*

    Dilated veins here form internal hemorrhoids.

4. *Lymphatics (297)—to sacral and internal iliac nodes*

5. *Nerves (306)*

    i. sympathetic (L1–L2)—inferior hypogastric (pelvic) plexus

    ii. parasympathetic (S2–S4)—pelvic splanchnics

B. Below the Line—that portion derived from proctoderm (ectoderm). The epithelium is squamous.

1. *Arteries (373)—inferior rectal arteries from the pudendal*

2. *Veins (365, 370)—The inferior rectal veins form an external venous plexus. They enter the pudendal veins and then the internal iliacs (systemic veins).*

    Dilated veins here form external hemorrhoids.

3. *Lymphatics (297)—to the inguinal nodes*

4. *Nerves (306)—somatic innervation—inferior rectal nerves from the pudendal (have pain fibers)*

II. ANAL MUSCLES (366, 367)—sphincters

A. Internal Anal Sphincter—smooth muscle continuation of circular muscle layer (innervated by the pelvic splanchnic nerves)

B. External Anal Sphincter—voluntary (innervated by rectal branches of the pudendal nerve). Some of the fibers of the external sphincter are connected with fibers from the puborectalis muscle.

In traumatic conditions involving the anal canal (e.g., 4th-degree lacerations during childbirth) or surgical procedures which include the rectal mucosa (e.g., proctoepisiotomy), in order to avoid future fecal incontinence, the fine muscle layers surrounding the anal mucosa should be approximated. The superficial transverse perineal muscle and its fascial covering (capsule) should also be approximated.

III. SPACES (368)

A. Intersphincteric Space—located between the internal and external sphincters—contains anal intermuscular glands and terminal longitudinal muscle fibers from the rectum

An infection in the glands may produce an intersphincteric abscess which may spread vertically between the sphincters (superiorly to the supralevator space or inferiorly to the perianal area beneath the skin) or it may spread horizontally across the voluntary sphincter into the ischiorectal fossa.

B. Supralevator (Pelvirectal) Space—between the rectum and levator ani

C. Ischioanal (Ischiorectal) Fossa—external to the external sphincter

IV. OTHER LINES (365)

A. Anorectal Line—located ½ inch above the pectinate line. This area is a **transition zone** with columnar epithelium above and squamous epithelium below.

B. Anal Pectin—a white band below the pectinate line

C. Intersphincteric (Intermuscular) Groove—at the lower border of the internal sphincter

D. Hilton's White Line—a white to light blue band in the anal canal over the intersphincteric groove (depression) which marks the transition between internal and external sphincter predominance

Upper Limb

Chapter 7

Page numbers in black. Netter plate numbers in green.

Upper Limb—General

I. ARTERIES (442)

A. Axillary—continuous with the subclavian above and the brachial below

B. Brachial—supplies the anterior aspects of the arm

1. *Profunda brachii (Deep brachial)—supplies the posterior compartment*

2. *Radial—supplies the muscles on the lateral side of the forearm and forms the deep volar arch of the hand*

3. *Ulnar—supplies the flexors of the fingers and contributes to the palmar arches in the hand*

II. TRIBUTARIES OF THE AXILLARY VEIN (448, 449, 452)

A. Superficial Tributaries—begin in arches in the palm and back of the hand which then enter the basilic vein or cephalic vein near the wrist. The cephalic extends along the radial side of the forearm. The basilic extends up the ulnar side. The median cubital connects the cephalic to the basilic in the cubital fossa.

B. Deep Tributaries—accompany the deep arteries

III. LYMPH NODES (169)—The apical axillary group receives some efferents from the other 5 axillary groups. The lateral axillary group receives many of the lymphatics of the upper limb.

IV. MUSCLE INNERVATION (442, 443)

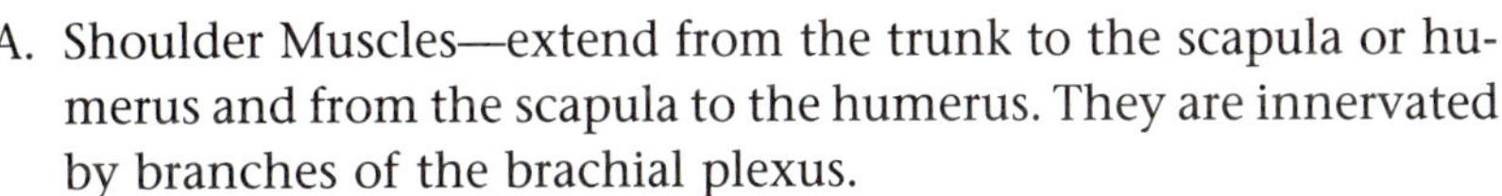

A. Shoulder Muscles—extend from the trunk to the scapula or humerus and from the scapula to the humerus. They are innervated by branches of the brachial plexus.

B. Arm

1. *Anterior compartment—innervated by the musculocutaneous nerve*

2. *Posterior compartment—innervated by the radial nerve*

C. Forearm

1. *Anterior compartment—innervated by the median and ulnar nerves*

2. *Posterior compartment—innervated by the radial nerve*

D. Hand—The ulnar nerve innervates muscles of the hand except for 2 lumbricals and 3 thenars which are innervated by the median nerve.

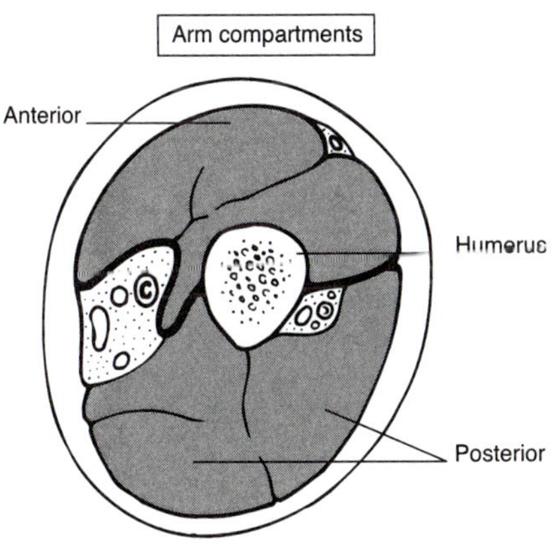

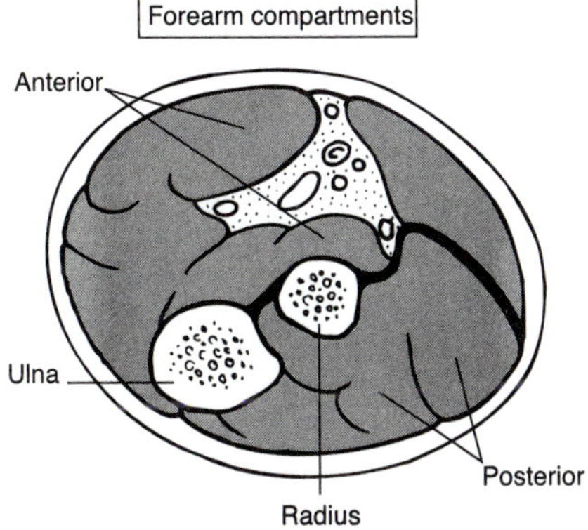

V. COMPARTMENT SYNDROMES—An increase in pressure within a closed space results in compromised circulation. Limb compartments are groups of muscles surrounded by rigid fibrous tissue. The *forearm and leg* are most commonly involved. Symptoms include pain, paresthesias, and paralysis.

VI. MUSCLE WEAKNESS FROM NERVE INJURIES

TABLE 7-1.

NERVE INJURY	COMMON NAMES	WEAKNESS
Long thoracic	Winged scapula	Serratus anterior
Thoracodorsal		Latissimus dorsi
Brachial plexus (Superior trunk C5–C6)	Duchenne's or Erb's paralysis	Deltoid (shoulder abductor) Abduction Forearm flexion Arm rotation
Brachial plexus (Middle trunk C7) Posterior cord	Radial nerve paralysis Cruch palsy	Extensors of arm, forearm, and wrist
Brachial plexus (Inferior trunk C8-T1)	Klumpke's paralysis	Wrist flexors Adduction and abduction of the fingers Digital flexors
Musculocutaneous		Biceps and brachialis
Median	Carpal tunnel	Flexion of proximal IP joints
Radial	Wrist drop	Forearm, wrist, and digit extension
Ulnar	Claw hand	Flexion and adduction at the wrist Inability to flex the 4th and 5th digits at the distal IP joints Weakness in abduction of the little finger and thumb

Shoulder

I. BONES (392, 393)

A. Scapula—a triangular-shaped bone on the dorsal thorax. It is located over ribs 2–7.

1. *Head*

   i. glenoid fossa (cavity)—for the head of the humerus. The glenoid labrum runs along the rim.

   ii. coracoid process—from the superior border of the head

2. *Neck—located medial to the glenoid cavity*

3. *Body—thin—constitutes the greater portion of the scapula*

4. *Acromion—extends from the lateral end of the spine—for the deltoid muscle*

5. *Surfaces*

   i. anterior (costal)—contains the subscapular fossa

   ii. posterior surface (dorsal)—contains:

      a. spine—a lateral projection—ends in the acromion

      b. supraspinous area—for the levator scapulae and supraspinatous

      c. infraspinous area

         1) axillary border—for teres major and minor

         2) vertebral border—for rhomboids

         3) inferior border—for slips of latissimus

6. *Angles—levels*

   i. superior angle of scapula—located at the level of T2

   ii. inferior angle of scapula—located at the level of T7

B. Clavicle (391) (sometimes referred to as the "strut of the shoulder")

1. *From the manubrium to the acromion*

2. *Connects the upper limb and trunk*

C. Head of the Humerus

II. SHOULDER INJURIES

A. Scapula

1. ***Fracture***

2. ***Dislocation of the glenohumeral joint ("shoulder"). The dislocation is due to the lack of muscles anteriorly and the shallowness of the glenoid fossa. The axillary nerve may be injured.***

B. Clavicle

1. ***Acromioclavicular ("shoulder") separation***

2. ***Fracture clavicle—a very common fracture which usually occurs at the junction of the middle and outer third of the bone.***

   The medial fragment is displaced upward and the lateral portion downward.

➢ III. MUSCLES (395)

A. Anterior Pectoral Muscles—innervated by the medial and lateral thoracic nerves

1. *Pectoralis major—adducts and medially rotates the humerus*
   - i. clavicular head—extends from the medial portion of the clavicle to the intertubercular groove of the humerus—flexes the humerus
   - ii. sternocostal head—from the sternum and ribs to the intertubercular groove of the humerus—extends the humerus
2. *Pectoralis minor—innervated by the medial and lateral thoracic (pectoral) nerves*
   - i. from 3rd, 4th, and 5th ribs to the medial border of the coracoid process of the scapula
   - ii. lies under the pectoralis major
   - iii. stabilizes, depresses, and rotates the scapula
   - iv. raises 3rd and 4th ribs in forced inspiration when the scapula is fixed
3. *Serratus anterior—innervated by the long thoracic nerve*
   - i. from the ribs to the medial border of the scapula
   - ii. draws the scapula forward over the thoracic wall and aids in raising the arm above the shoulder
4. *Subclavius—from the 1st rib to the clavicle—stabilizes and rotates the scapula and aids in lowering the pectoral girdle*

B. Anterior Muscle—Teres major (397)—innervated by the subscapular nerve (C5–C6)—from the infraspinous fossa of the scapula to the intertubercular groove of the humerus—adducts and internally (medially) rotates the humerus

C. Posterior Muscles (178, 395)

1. *From vertebral column to scapula*

   i. trapezius—innervated by the spinal accessory nerve (C3–C4)—from the occiput and nuchal ligament (spines of C7-T12) to the clavicle and scapular spine—elevates and retracts the scapula and raises the shoulder

   ii. levator scapulae—from the transverse processes of the vertebrae to the superior angle of the scapula—innervated by the dorsal scapular nerve (C5)—elevates the scapula

   iii. rhomboids—from the vertebrae to the medial aspect of the scapula—innervated by the dorsoscapular nerve (C5)—brings the scapula toward the vertebral column

2. *From vertebral column to humerus*

   i. latissimus dorsi—from the spines of the lower 6 thoracic vertebrae to the humerus—extends, adducts, and medially rotates the humerus—innervated by the thoracodorsal nerve (C6–C8)

D. Superior Muscle—Deltoid (395)—from the scapula and clavicle to the deltoid tuberosity of the humerus—**abducts** and rotates the humerus—innervated by the axillary (circumflex) nerve (C5–C6)

E. Posterior Scapular Muscles (178)—supraspinatous, infraspinatous, and teres major and minor

IV. MOVEMENTS OF THE SCAPULA

A. Upward—levator scapulae, rhomboids, and the upper part of the trapezius

B. Downward (Depression)—pectorals, latissimus dorsi, and the lower part of the trapezius

C. Forward (Over the Thoracic Wall) (Protraction)—pectorals and serratus anterior

D. Backward—rhomboids, trapezius, and latissimus dorsi

V. ROTATOR CUFF (396)—extends from the scapula to the humerus and strengthens the shoulder joint by holding the head of the humerus in the glenoid cavity. The cuff is musculotendinous.

A. Supraspinatous—from the supraspinous fossa to the greater tubercle—innervated by the suprascapular nerve (C4–C5)

B. Infraspinatous—from the infraspinous fossa to the tubercle of the humerus—innervated by C5

C. Teres Minor—from the lateral aspect of the scapula to tubercle of the humerus—innervated by the axillary nerve (C5–C6)

D. Subscapularis—from the subscapular fossa to the lesser tubercle of the humerus—innervated by the subscapular nerve (C5–C6)

➢ VI. SCAPULOCLAVICULAR LIGAMENTS (394, 396, 402)

A. Coracoacromial—located between the coracoid process and the acromion—anterior to the joint—prevents superior dislocation of the head of the humerus

B. Acromioclavicular—between the acromion and the clavicle—located above and below the acromioclavicular joint

C. Trapezoid—between the coracoid process and the clavicle

➢ VII. CAPSULE (394, 397)—surrounds the shoulder joint

A. Fibrous (ligamentous)

1. *Glenohumeral ligament—attached to the glenoid cavity, the neck of the humerus, and the coracoid process*

2. *Transverse humeral ligament—from the greater to the lesser tubercle of the humerus—passes over the intertubercular groove which transmits the tendon of the long head of the biceps brachii*

3. *Coracohumeral—coracoid process to the greater tubercle of the humerus—located at the superior portion of the capsule*

B. Synovial—lines the capsule and forms the sheath for the biceps brachii

C. Muscles and Tendons around the capsule

1. *Superior—supraspinatous tendon (blends into the capsule)*

2. *Inferior—triceps brachii*

3. *Posterior—teres minor and infraspinatous*

4. *Anterior—subscapularis tendon*

VIII. JOINTS (391, 394)

A. Sternoclavicular—synovial. The articular disk produces distinct spaces within the joint.

B. Acromioclavicular—synovial

It may be palpated at the lateral end of the clavicle.

C. Glenohumeral (Shoulder Joint)—located between the glenoid fossa of the scapula and the head of the humerus—synovial ball and socket joint—has gliding, rolling, and rotation movements

Its anteroinferior aspect lacks good support.

IX. BURSAE (394)

A. Subacromial—allows movement between the scapula and the thoracic wall—located between the acromium, deltoid muscle, and the head of the humerus

1. *Subdeltoid—that portion of the subacromial bursa located under the deltoid muscle*

B. Subscapular—communicates with the cavity of the shoulder joint—located between the neck of the scapula and the subscapularis muscle

C. Latissimus Dorsi Bursa—between the tendons of the latissimus dorsi and teres major

X. SHOULDER MOVEMENTS (397)

A. Adduction of the Humerus (395)

1. *Latissimus dorsi—from the lumbar aponeurosis, ribs, and ilium to the intertubercular groove of the humerus*
2. *Pectoralis major—from the medial clavicle and sternum to the greater tubercle of the humerus*

B. Abduction of the Humerus (178)

1. *Deltoid—from the lateral clavicle and spine of the scapula to the deltoid prominence (tuberosity) of the humerus*
2. *Supraspinatous—from the supraspinous fossa to the tubercle of the humerus*
3. *Infraspinatous—from the infraspinous fossa to the greater tubercle of the humerus and the capsule of the shoulder joint*
4. *Teres major—that portion of the muscle which extends from the inferior angle of the scapula to the intertubercular groove of the humerus*
5. *Subscapularis—from the subscapular fossa to the lesser tubercle of the humerus and the capsule of the shoulder joint*

- C. Raise the Shoulder—Trapezius (395)—from the occipital bone and spinous processes of C7–C12 to the scapula, acromion, and clavicle
- D. Pull the Shoulder Down and Forward (174)—Pectoralis minor
- E. Take the Scapula over the Thoracic Wall—Serratus anterior—from ribs to scapula
- F. Draw the Arm Across the Chest—Pectoralis major—from the medial clavicle and sternum to the greater tubercle of the humerus

XI. MUSCLES WHICH STABILIZE THE SHOULDER (395)

A. Extend from the Trunk to the Shoulder Girdle and hold the shoulder to the trunk

1. *Trapezius—from the occipital bone and the spinous processes of C7-T12 downward to the clavicle and scapula*
2. *Pectoralis minor—from the ribs to the coracoid process of the scapula*
3. *Rhomboids—from the spinous processes of T2–T5 to the clavicle and scapula*
4. *Levator scapulae—from vertebrae C1–C4 to the medial aspect of the scapula*
5. *Serratus anterior—from the ribs to the medial border of the scapula*
6. *Rotator cuff muscles (Supraspinatous, Infraspinatous, Teres minor, and Subscapularis)—from the scapula to the humerus*

B. Extend from the Shoulder Girdle to the Humerus and hold the head of the humerus against the glenoid cavity

1. *Rotator cuff muscles*
2. *Teres major—from the scapula to the lesser tubercle of the humerus*
3. *Deltoid—from the lateral clavicle and spine of the scapula to the deltoid prominence of the humerus*

C. Extend from the Trunk to the upper Humerus and stabilize the shoulder in movements of the forearm and hand

1. *Latissimus dorsi—from the lumbar aponeuroses of the ribs and ilium to the intertubercular groove of the humerus*

2. *Pectoralis major*

   i. clavicular head—from the medial portion of the clavicle to the intertubercular groove of the humerus

   ii. sternocostal head—from the sternum and ribs to the intertubercular groove of the humerus

XII. ARTERIES (398)

A. Axillary

1. *The thoracoacromial (acromiothoracic) has branches to the clavicle, pectoralis major, and deltoid muscles.*
2. *The lateral thoracic to pectoral muscles*
3. *The subscapular branch to the teres major, latissimus dorsi, and serratus anterior muscles*
4. *The circumflex humeral branch to the humerus*

B. Suprascapular (Transverse scapular)—from the thyrocervical trunk of the subclavian—supplies the upper scapula

C. **Anastomoses**—The subscapular (from the axillary) forms anastomoses with scapular branches of the thyrocervical trunk.

XIII. VEINS (399)—axillary

XIV. LYMPHATICS (452)—to axillary nodes

XV. NERVES TO THE SHOULDER JOINT (400)—branches of the subscapular, lateral pectoral, and axillary nerves

Axilla

I. THE AXILLA (400)—represents the space between the superolateral chest wall and the upper limb.

A. Contains the axillary artery, the axillary vein, and the axillary portion of the brachial plexus. The axillary portions begin at the lateral border of the 1st rib and extend to the lower border of the teres major muscle. The arteries, veins, and nerves are covered by deep fascia continuous with the prevertebral fascia.

B. Contains the axillary nodes

C. The nerve, artery, and vein to the humerus go around the surgical neck of the humerus.

II. WALLS (399)

A. Anterior—pectoral muscles

B. Medial—5 upper ribs

C. Lateral—coracobrachialis and biceps brachii muscles, proximal humerus

D. Posterior—subscapularis. The quadrangular space is located in the posterior wall. It contains the posterior circumflex humeral vessels and the axillary nerve.

E. Floor—skin and axillary fascia

III. FASCIA (399)

A. Coracoclavicular (Clavipectoral)—extends from the clavicle to the axillary fascia

B. Axillary Sheath—a continuation of prevertebral fascia—encloses axillary vessels and nerves

IV. AXILLARY ARTERY (398, 405)—The portion of an artery located between the subclavian and the brachial. It begins at the lateral border of the 1st rib, passes posterior to the pectoralis minor, and ends at the insertion of the teres major on the humerus to become the brachial artery.

A. Superior Thoracic—supplies the upper part of the serratus anterior and sometimes the pectoralis major and minor

B. Thoracoacromial (Acromiothoracic)—supplies the pectorals, the deltoid muscle, and the acromioclavicular joint

C. Lateral Thoracic—supplies lateral half of the breast, the pectoral, and portions of the serratus anterior muscles

D. Subscapular—supplies the posterior axillary muscles and the scapula—**anastomoses** with the transverse cervical artery

E. Circumflex Humeral Arteries—supply the humerus, the shoulder joint, and the attached muscles

V. AXILLARY VEIN (175)—a continuation of the basilic vein. It has superficial and deep tributaries (both have valves). The axillary vein lies medial to the artery.

A. Receives veins accompanying the corresponding arteries

B. Receives the cephalic vein which drains the superficial lateral aspects of the upper limb. A portion of the cephalic vein is located between the deltoid and pectoralis major.

VI. LYMPH NODES (169)

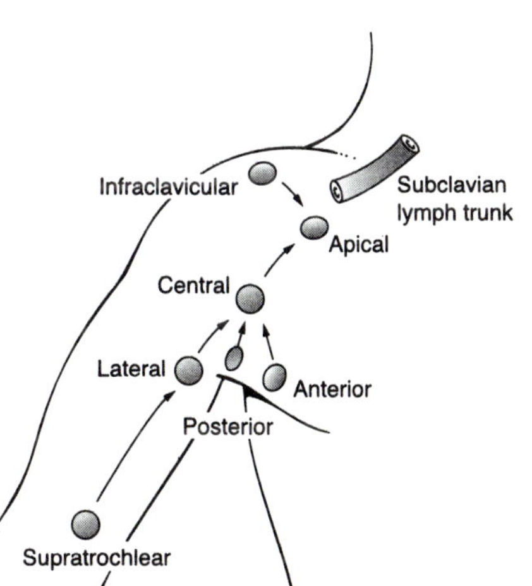

A. Apical Group—located near the axillary vein—receive efferents of the other 5 groups

B. Anterior (Pectoral) Group—receive efferents from the anterolateral wall down to the umbilicus and the lateral half of the breast. They are located near the lower border of the pectoralis muscle and adjacent to the serratus anterior.

C. Lateral Group—receive efferents from the deep tissues of the upper limb and superficial aspects of the medial side of the upper limb. They are located along the axillary vein.

D. Posterior (Subscapular) Group—receive efferents from the posterior trunk. They are located in the posterior axillary fold and on the subscapularis muscle. They drain the posterior chest and receive some anterior pectoral lymphatics.

E. Central Group—receive efferents from anterior, posterior, and lateral groups. They are located low in the axilla.

F. Infraclavicular Group—receive efferents from the superficial parts of the upper limb

VII. INJURIES TO NERVES IN THE AXILLA

A. **Axillary Nerve (C5–C6) (400)—Dislocation of the shoulder may damage the axillary nerve. Injury near the surgical neck of the humerus may affect the teres minor muscle (lateral humeral rotation).**

B. **Long Thoracic Nerve (C5–C7) (177)—Injury affects the serratus anterior muscle (movement of the scapula over the thoracic wall) ("Winged Scapula").**

C. **Musculocutaneous Nerve (C5–C7) (443)—Injury affects the biceps and brachialis (decreased flexion at the elbow)**

D. **Thoracodorsal Nerve (C5–C7) (397)—Surgical injury may produce paralysis of the latissimus dorsi (adduction and medial rotation of the humerus).**

Arteries to the Upper Limb

I. AXILLARY ARTERY (405)

A. Course—The axillary artery begins at the lateral border of the 1st rib, passes posterior to the pectoralis minor, and ends at the insertion of the teres major on the humerus. Its course follows an approximate line from the midpoint of the clavicle to the lateral ⅓ of the anterior axillary fold.

B. Branches in the Axilla—superior thoracic, a small branch to the thoracic wall muscles—acromiothoracic to the acromion, clavicle, deltoid, and pectoralis major and minor—lateral thoracic to the lateral thoracic wall—subscapular to the lateral thoracic wall (teres major, latissimus dorsi, and serratus anterior)

C. Branches in the Arm—posterior circumflex humerals to the upper humerus, shoulder joint, triceps brachii, deltoid, and teres muscles

II. BRACHIAL ARTERY (404, 405)

A. Course—The brachial artery begins at the tendon of the teres major and ends in the forearm, near the neck of the radius, by dividing into the radial and ulnar arteries. In the lower portion of the arm, it accompanies the median nerve.

B. Relationships

1. *In the arm—The brachial artery lies on the medial side of the anterior muscular compartment.*
2. *In the antecubital fossa*
   - i. anterior—veins
   - ii. posterior—brachialis muscle
   - iii. medial—median nerve
   - iv. lateral—tendon of biceps brachii

C. Branches

1. *Deep brachial (Profunda brachii)—In the posterior compartment, it follows the radial nerve in the radial groove and gives off recurrent, collateral, and muscular branches (to the brachialis and deltoid).*
2. *Ulnar collaterals—superior and inferior*
3. *Muscular (including one to the biceps) and humeral nutrient branches*

III. ULNAR ARTERY (405, 417)

A. Course—The ulnar artery begins below the bend of the elbow (close to the radial neck) and goes down the forearm (in the anterior compartment) to the radial side of the pisiform bone.

B. Branches

1. *Recurrent—to forearm muscles*
2. *Muscular—to ulnar side muscles*
3. *Interosseus*
4. *Carpal—to carpals and* ***anastomoses*** *with radial branches*
5. *Volar arch—curves across the palm convex downward and has branches to volar digital arteries*

IV. RADIAL ARTERY (405, 417)

A. Course

1. *The radial artery begins at the bifurcation of the brachial below the elbow bend and continues along the radial side of the forearm (in the anterior compartment), around the lateral carpals into the palm, to unite with the volar (a branch of the ulnar) to form the deep volar arch.*
2. *In the wrist, it is palpable over the distal radius (lateral to the flexor carpi radialis tendon and the pisiform bone).*

B. Branches—supply the radial portions of the forearm, wrist, and hand

1. *Muscular—on the radial sides of the forearm and hand*
2. *Branches to carpals and metacarpals*

V. ARTERIAL ARCHES (435)

A. Palmar Arches

1. *Superficial—mostly from the ulnar artery— gives off digital arteries*
2. *Deep—mostly from the radial artery. The deep arch enters the palm of the hand and forms metacarpal arteries. It has* ***anastomoses*** *with the digital arteries of the superficial arch.*

B. Dorsal Arches

1. *From the radial and ulnar arteries to dorsal arteries of the thumb and index finger*
2. *The dorsal metacarpal arteries branch into the dorsal digital arteries.*

Brachial Plexus

I. ORIGINS (ROOTS) (401)—formed by the union of the ventral anterior rami of nerves C5-T1 (in the neck). Anterior and posterior roots exit the intervertebral foramen to unite and form a spinal nerve trunk. The roots form 3 trunks.

A. The dorsal scapular nerve (C5)

B. The long thoracic nerve (C5–C7) innervates the serratus anterior muscle.

II. TRUNKS (401)—located in the posterior triangle of the neck. Each of the three trunks divides into anterior and posterior divisions posterior to the clavicle. The 3 trunks form 6 divisions.

A. Superior Trunk (C5–C6)

Severe injury to the superior trunk (Erb-Duchenne Paralysis) may result from breech delivery childbirth (upward traction of the upper limb) causing "Headwaiters Tip Hand" in which the hand is to the side in internal rotation, the fingers and wrist are flexed, and lateral rotation and abduction are lost at the shoulder. The involved muscles and their impaired function include:

1) Deltoid—decreased abduction and rotation of the humerus
2) Supraspinatous and infraspinatous—paralysis of flexors at the elbow
3) Biceps, Brachioradialis, and Brachialis—impaired medial rotation of the arm
4) Pectoralis major, Teres major, Latissimus dorsi
5) Subscapularis—weakened adductors and medial rotators of the humerus

1. *Suprascapular branch to spinatus muscles*
2. *Subclavius branch*

B. Middle Trunk (C7)

Injury to the middle trunk may result in radial nerve palsy and "wrist drop" and weakness of the brachioradialis muscle.

C. Inferior (Lower) Trunk (C8-T1)

Injury to the inferior trunk from severe upward traction of the shoulder may result in Klumpke's Paralysis. Paralysis of the interossei results in decreased abduction and adduction of the fingers plus decreased wrist flexion due to paralysis of the flexors of the digits.

III. DIVISIONS (401)—located in the axilla behind the clavicle

A. Anterior (3)—supply the flexors of the upper limb

B. Posterior (3)—supply the extensors of the upper limb

IV. CORDS (401)—located in the axilla and named according to their relationship to the axillary artery

A. Posterior Cord—formed by the union of all of the posterior divisions. It gives off the 3 subscapular nerves and then divides into the axillary and radial nerves.

1. *Upper subscapular (C5–C6)—innervates the subscapularis muscle*
2. *Middle subscapular (Thoracodorsal) (C6–C8)—innervates the latissimus dorsi*

   Injury in the medial axilla may result in decreased adduction and medial rotation of the arm.

3. *Lower subscapular (C5–C7)—innervates the subscapularis and teres major*
4. *Axillary (C5–C6)—innervates:*
   - i. the teres minor and deltoid muscles
   - ii. the shoulder joint
5. *Radial (C5–C7)—innervates:*
   - i. triceps and brachialis
   - ii. the extensors of the wrist, hand, and fingers
   - iii. the elbow and wrist joints

B. Lateral Cord—formed by the union of the anterior divisions of the upper and middle trunks. It divides into the musculocutaneous nerve and the lateral root of the median nerve.

1. *Lateral pectoral (C5–C7)—innervates:*
   - i. the pectoral muscles
   - ii. the shoulder joint
2. *Musculocutaneous (C5–C7)—innervates the biceps brachii, brachialis, and coracobrachialis (which it pierces)*

   Injury in the axilla may weaken the biceps and brachialis and decrease flexion at the elbow.

3. *Median (C6-T1)—innervates:*

   i. the pronators and the lateral flexors of the wrist and digits

   ii. the lateral lumbricals

   iii. the wrist joint and the metacarpophalangeal (MP) and interphalangeal (IP) joints of the hand

C. Medial Cord—the continuation of the anterior division of the inferior trunk. It divides into the ulnar nerve and the lateral root of the median nerve.

1. *Medial pectoral (Anterior thoracic)—innervates the pectoral muscles*

2. *Ulnar (C8 and T1)—the continuation of the medial cord—innervates:*

   i. medial flexors of the wrist and fingers

   ii. medial lumbricals

   iii. flexor, abductor, and opponens digiti minimi

   iv. palmar and dorsal interossei muscles

   v. the wrist and MP and IP joints

V. COURSE OF NERVES (400, 442–447)

A. Axillary Nerve (C5–C6)—accompanies the posterior circumflex humeral artery

1. *In the axilla*

   Dislocation of the shoulder may damage the axillary nerve.

2. *In the arm—lies inferior to the capsule of the shoulder joint and medial to the neck of the humerus*

   Fracture of the surgical neck of the humerus may damage the axillary nerve and cause weakness in the deltoid and teres minor resulting in decreased ability to abduct and rotate the humerus.

B. Median Nerve (C6–C8 and T1) (444)—formed below the pectoralis minor by a medial and a lateral cord of the brachial plexus

1. *In the axilla—lateral and superficial to the axillary artery*

2. *In the arm—lateral to the brachial artery and then crosses over to the medial side of the artery—no branches in the arm*

   Severance of the median nerve in the arm affects the anterior forearm compartment muscles and results in paralysis of forearm pronation and inability to flex the index and middle fingers.

3. *In the elbow—medial to the brachial artery and under the pronator teres muscle*
4. *In the forearm—lies between the heads of the pronator teres and then gives off the anterior interosseus branch which innervates the deep anterior forearm muscles. It supplies the superficial anterior forearm muscles (it generally supplies the flexors and pronators on the radial side of the forearm).*
5. *Within the wrist—lies between the tendons of the flexor muscles and passes thru the carpal tunnel*

   Severance of the median nerve at the wrist results in inability to flex and pronate thenar muscles and fully flex the index and middle fingers.

   Compression at the wrist due to swelling results in Carpal Tunnel Syndrome (CTS).
6. *In the hand—The median nerve supplies the short muscles of the thumb, the lateral lumbricals, and the skin of the central part of the palm.*

C. Radial Nerve (C5-T1)—a continuation of the posterior cord

1. *In the axilla—The nerve lies between the triceps muscle and the medial aspect of the arm.*
2. *In the arm (446)—The radial nerve enters posterior to the radial artery and passes down the spiral (radial) groove.*

   It may be injured in a fracture of the middle of the humerus. Nerve injury due to fracture of the middle of the humerus may cause "wrist drop."

   The radial nerve passes down between the head of the triceps brachii and at the lateral epicondyle of the humerus. It divides into superficial and deep branches.

   i. The posterior brachial cutaneous branch supplies the skin at the back of the arm.

   ii. muscular branches to the triceps and aconeus
3. *In the elbow—The cubital fossa contains part of the radial nerve. At the end of the humerus and lateral to the medial epicondyle, it divides into superficial and deep branches.*
4. *In the forearm (447)*

   i. the deep (posterior interosseus) (muscular) branch—supplies a branch to the supinator muscle and then goes around the neck of the radius to supply the posterior compartment

   ii. The superficial (sensory) branch goes under the brachialis.

5. *In the hand—The radial nerve enters the posterior side to supply the dorsum (lateral aspect) of the hand and the lateral digits down to the proximal IP joints.*

D. Ulnar Nerve (C8-T1)—**The ulnar nerve is the most common peripheral nerve injured during surgical procedures.**

1. *In the arm—The ulnar nerve lies on the triceps muscle, medial to the brachial artery. Halfway down the arm, it passes from the anterior compartment to the posterior compartment. It continues on to pass between the medial epicondyle (posteriorly) and the olecranon.*

   Injury due to a fracture dislocation posterior to the medial epicondyle may result in decreased ability to flex the wrist and decreased flexion at the 4th and 5th IP joints.

➢ 2. *In the forearm (445)—The ulnar nerve enters the forearm by passing between the heads of the flexor carpi ulnaris and then becomes more superficial and gives off its dorsal cutaneous branch.*

➢ 3. *In the wrist (445)—It lies lateral to the tendon of the flexor carpi ulnaris and passes superficially to the flexor retinaculum.*

   Severance of the ulnar nerve at the wrist results in decreased flexion at the 4th and 5th IP joints

➢ 4. *In the hand (445)—The ulnar nerve lies superficial to the flexor retinaculum and then divides into superficial and deep branches. It supplies the skin over the 4th and 5th digits, the hypothenar eminence, and several short intrinsic muscles.*

VI. ANS IN THE UPPER LIMB—Sympathetic nerves are the primary ANS nerves to the upper limb. They come from the stellate ganglion (a fusion of the inferior cervical and 1st thoracic ganglion of the sympathetic cervical chain).

Arm

I. HUMERUS BONE (392, 393) —articulates with the scapula and the 2 bones of the forearm

A. Proximal End

1. *Head*

2. *Anatomical neck—has the intertubercular (bicipital) groove for the long tendon of the biceps*

   i. greater tubercle—for the supraspinatous, infraspinatous, and teres minor

   ii. lesser tubercle—for the subscapularis

   iii. intertubercular (bicipital) groove—lies between the greater and lesser trochanters—for the insertions of the teres major and minor, pectoralis major, latissimus dorsi, and long tendon of the biceps

B. Shaft—contains the spiral (radial) groove (for the profunda brachii artery and the radial nerve)

1. *The **surgical neck** is that part of the shaft located between the tubercles and the insertion of the pectoralis major and latissimus dorsi (i.e., between the head and shaft).*

2. *Deltoid tuberosity—on the upper portion of the shaft*

C. Distal End

1. *Medial epicondyle—for pronator and flexors of the forearm*

2. *Lateral epicondyle—for supinator and extensors*

3. *Trochlea—a pulley-shaped area on the medial side. The trochlea articulates with the trochlear notch of the ulna.*

4. *Olecranon fossa—for the olecranon of the ulna*

5. *Coronoid fossa—receives the coronoid process of the ulna*

6. *Capitulum—ball-shaped area on the lateral side—articulates with the head of the radius*

II. HUMERAL FRACTURES (393)

A. Proximal Humeral End

1. *Anatomical neck—fracture occurs within the capsule—frequently results in avascular necrosis of the head*

B. Shaft—fracture occurs outside the capsule. Displacement of the upper and lower portions of the shaft depend on whether the fracture is above or below the deltoid insertion.

1. ***Surgical neck—Displacement of the upper portion of the humerus is medial and upward (toward the axilla). The axillary nerve may be injured.***

2. ***Middle of the shaft (Midhumeral)—Fractures which include the radial (spiral) (musculospiral) groove may affect the radial nerve and thus the forearm extensors.***

C. Distal Humeral End Fractures

1. ***Supracondylar—Fractures may injure the median nerve and brachial artery (a significant risk).***

2. ***Medial epicondyle—Fractures may affect the ulnar nerve.***

3. ***Lateral condyle—from a fall on the hand***

4. ***Intercondylar—between the medial and lateral condyles***

➢ **III. FASCIA (406)—The deep fascia of the arm is continuous with the deep fascia of the forearm and continuous with the intermuscular septa (medial and lateral) which divide the arm and forearm into anterior and posterior compartments.**

➢ **IV. MUSCLES (406)**

➢ A. Anterior Compartment (402, 406)—innervated by musculocutaneous nerve (C5–C6). Muscles extend from the scapula and humerus to the humerus and the radius and ulna near the elbow.

1. *Coracobrachialis—from the coracoid process to the humerus—flexes and adducts the arm at the shoulder joint*

2. *Brachialis—from the anterior surface of the humerus to the ulna—flexes forearm in pronated position*

3. *Biceps brachii—from the scapula to the radius—supinates and flexes the forearm (a strong supinator)*

➢ B. Posterior Compartment (403, 404, 406)—The extensor group is innervated by radial nerve (C6–C8). The muscles extend from the scapula and humerus to the ulnar olecranon.

1. *Triceps brachii—extends the forearm*

   i. long head—scapula (below the glenoid cavity) to the olecranon

   ii. lateral head—spiral (radial) groove to the olecranon

2. *Anconeus—from the lateral epicondyle of the humerus to the ulna—extends the forearm*

V. MUSCLES WHICH MOVE THE HUMERUS—include the coracobrachialis, deltoid, latissimus dorsi, pectoralis major, supraspinatous, and infraspinatous

VI. COMPARTMENTS (406)

A. Compartmental Separation

1. *In the upper part of the arm—partially by muscles from the thorax*
2. *On the medial side of the arm—by the major blood vessels and several muscles*
3. *In the lower half of the arm—by intermuscular septa*

B. Structures Common to Both Compartments—radial and ulnar nerves, deep brachial (profunda), and collateral arteries

VII. BRACHIAL ARTERY (404, 405)

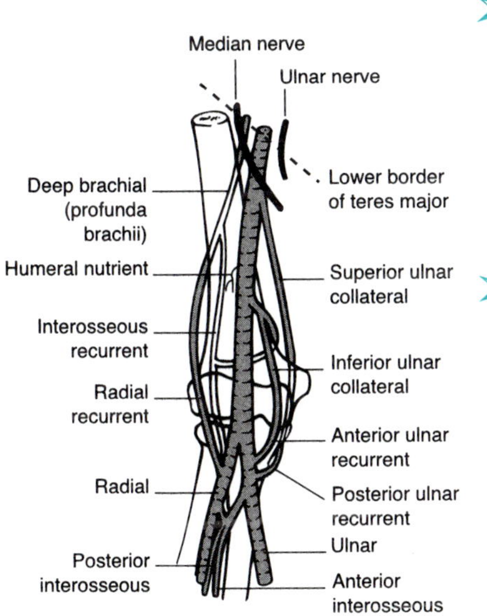

A. Course (404)—The brachial artery begins at the lower portion of the teres major muscle and descends on the medial side of the arm posterior to the biceps muscle and immediately lateral to the median nerve. It ends in the forearm by dividing into the radial and ulnar arteries. The radial nerve accompanies the brachial artery in the upper part of the arm. The brachial artery supplies the anterior aspects of the arm.

B. Branches (405)

1. *Humeral nutrient branches*
2. *Deep brachial artery (Profunda brachii)—passes down the spiral groove of the humerus and gives branches to the elbow. It supplies the posterior compartment.*
3. *Ulnar collateral arteries*
4. *Muscular branches (including one to the biceps)*
5. *Terminal branch—contributes to the formation of elbow* ***anastomoses***

VIII. VEINS (448) —(have valves)

A. Superficial—larger than the deep veins with which they communicate

1. *Basilic—extends from the back of the hand to the brachial vein. It courses anteromedially in the arm.*
2. *Cephalic—from the radial end of the dorsal venous arch of the hand to the axillary vein*

3. Median cubital—In the antecubital fossa, it connects the cephalic and basilic veins.

B. Deep—accompany arteries—begin in the hand as volar digital veins and end by draining into the axillary vein

IX. EPITROCHLEAR LYMPH NODES (452)—located around the medial epicondyle of the humerus

X. NERVES (443, 446)

A. Musculocutaneous (402, 406)—supplies the anterior compartment

B. Radial (403, 406, 446)—located posterior to the humerus, in the spiral groove, along with the profunda brachii artery

1. Motor—innervates the posterior compartment

Fractures of the distal humerus may damage the radial nerve.

2. Sensory—The posterior brachial branch supplies the skin of the back of the arm.

C. Ulnar (404, 406)—descends in the arm medial to the brachial artery and passes posterior to the medial epicondyle—no branches in the arm

D. Median (404, 406)—no branches in the arm but it may be palpated as it crosses the brachial artery in the middle portion of the arm (i.e., approximately ½ of the way down)

Severance of the median nerve in the arm produces paralysis which includes the muscles of the anterior forearm compartment, loss of forearm pronation, and the inability to flex the index and middle fingers.

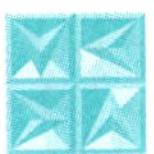

Elbow

The elbow is a synovial hinge joint with two articulations (humeroulnar and humeroradial).

I. BONES (407)

A. Trochlea—articulates with the trochlear notch of the ulna

B. Capitulum of Humerus—articulates with the head of the radius

II. LIGAMENTS (408)—form one large capsule

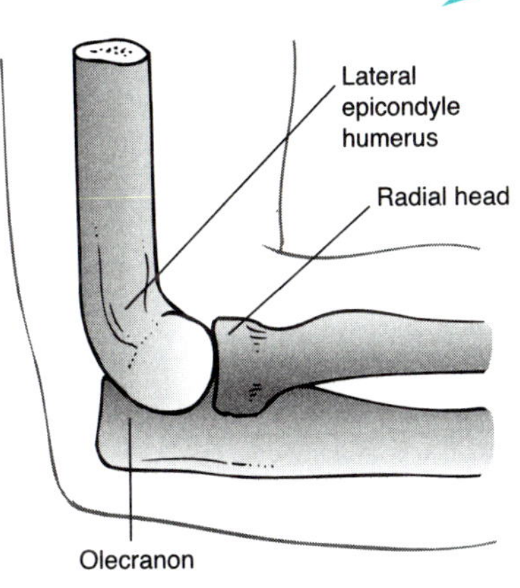

A. Anterior—from ridges in front of the epicondyles to the coronoid process

B. Radial Collateral (Lateral)—from the lateral epicondyle of the humerus to the annular ligament of the radius and the radial notch

C. Ulnar Collateral (Medial)—from the medial epicondyle and trochlea of the humerus to the coronoid process of the ulna and medial edge of the olecranon

D. Posterior—Decussating fibers extend from the humerus to the olecranon.

III. DISLOCATIONS (408)—subluxation of the head of the radius over the annular ligament—due to pulling the extended limb

IV. BURSAE—The synovial layer lines the fibrous capsule and extends around the neck of the humerus.

A. Radioulnar—between the supinator and the radiohumeral joint

B. Olecranon—between the olecranon process, triceps tendon, and skin

V. ANTECUBITAL FOSSA (404)

A. Borders—Sides: epicondyle of humerus, brachioradialis, and pronator teres
Floor: brachialis and supinator

B. Contents—brachial artery and veins, radial and ulnar arteries, median nerve, and tendon of biceps brachii

VI. ARTERIES (405)—The radial, ulnar, and profunda brachii arteries provide anastomoses via collateral and recurrent arteries. There are both anterior and posterior anastomoses.

VII. CUBITAL (EPITROCHLEAR) LYMPH NODES (452)—efferents to the axillary nodes

VIII. NERVES—articular branches of radial, ulnar, and median nerves.

The ulnar nerve runs and may be palpated in the medial olecranon groove. It may be damaged in dislocations or fractures of the elbow, causing paralysis to the flexor carpi ulnaris and radial deviation of the hand.

Forearm

The forearm is located below the elbow and above the wrist.

➢ **I. BONES (409, 420, 421)—joined by a fibrous interosseous membrane.**

A. Radius—rotates around the head of the ulna

1. *Proximal portion*

i. head—cylindrical—has a concave upper surface. The head is palpable below the lateral epicondyle of the humerus.

a. receives the capitulum of the humerus

b. held in place by the annular ligament

ii. neck

iii. tuberosity—distal to the neck—for the biceps brachii tendon

2. *Body*

i. anterior oblique line

ii. interosseous border—for the interosseous membrane

3. *Distal portion*

i. ulnar notch—for the head of the ulna

ii. styloid process—located on the lateral surface

a. for the brachioradialis muscle and the external collateral ligament of the wrist

b. extends farther distally than the ulnar styloid process

iii. dorsal tubercle—for the dorsal carpal ligament

B. Ulna—longer than the radius

1. *Proximal portion*

i. olecranon—clasps the distal end of the humerus

ii. coronoid process

The coronoid process may be dislocated posteriorly due to a fall on the hyperextended arm.

a. radial notch for the head of the radius

b. tuberosity for the brachialis muscle

iii. supinator fossa—attachment for supinator

iv. trochlear notch—formed by the olecranon and the coronoid process—articulates with the trochlea

2. *Body—attachment for the interosseous membrane*

3. *Distal portion*

i. head—articulates with the disk of the wrist joint

ii. styloid process—for the ulnar collateral ligament

iii. articular surface—articulates with the radius. The articular disk in the distal radioulnar joint produces distinct spaces within the joint.

II. FOREARM FRACTURES

A. Radius

1. *Radial head—due to a fall on the elbow*
2. *Fractures of the head and neck of the radius may also be due to a fall on the palm of the hand.*
3. *Colles' fracture—a fracture of the lower end of the radius, frequently from a fall on the extended hand*

i. Displacement of the proximal portion is anteriorly.

ii. Displacement of the distal portion is posteriorly. This may form a prominence on the back of the wrist and medial displacement of the hand.

B. Ulna

1. *May occur in combination with a fracture or dislocation of the radius. A fracture of the ulnar shaft along with a dislocation of the head of the radius is called "Monteggia's fracture."*
2. *Olecranon process—fracture from trauma*

III. RADIOULNAR JOINTS (409)—located near the proximal and distal ends of the bones—involved in pronation and supination

A. Proximal—the head of the radius with the radial notch of the ulna

B. Distal—the head of the ulna with the ulnar notch of the radial head

IV. MUSCLES (410–419)—Forearm muscles are sometimes classified into a flexor-pronator group, extending from the medial epicondyle of the humerus; a deep flexor group, extending from the anterior aspects of the radius and ulna; an extensor-supinator group from the lateral epicondyle of the humerus; and a deep extensor group from the posterior aspects of the radius and ulna.

A. Anterior Compartment (406)—mostly innervated by the **median nerve**

1. *Pronating of the forearm (410)*

i. pronator quadratus—from the anterior ulna to the distal radius

ii. pronator teres—from the humerus to the shaft of the radius. It is also a flexor of the forearm.

2. *Flexing of the hand or fingers—generally innervated by the median nerve*

➢ i. superficial flexors (416)—generally arise by a common tendon from the medial epicondyle of the humerus to insert on the bones of the hand

a. flexor carpi radialis—from the medial epicondyle of the humerus to the 2nd and 3rd metacarpals

b. palmaris longus—from the medial epicondyle to the flexor retinaculum and palmar fascia

c. flexor carpi ulnaris—from the humerus and ulna to the hamate bone—innervated by the **ulnar** nerve

d. flexor digitorum superficialis—from the humerus to the middle digital phalanges

➢ ii. deep flexors (418)—arise from the volar aspect of the forearm bones

a. flexor digitorum profundus—from the ulna to the distal phalanges—flexes the terminal phalanges

b. flexor pollicis longus—from the radius to the distal phalanx of the thumb—flexes the thumb

➢ B. Posterior Compartment(406)—extensors of the hand and fingers and supinators of the forearm—located on the back of the forearm—all innervated by the **radial nerve**

➢ 1. *Superficial group (414)—arise from the lateral epicondyle of the humerus (tendinous at the wrist) to insert on the bones of the hand and fingers. They are located in the extensor synovial sheaths and held in place by the extensor retinaculum.*

i. brachioradialis—to the styloid process of the radius. (It lies toward the front of the forearm and forms the lateral boundary of the cubital fossa; it acts as a forearm flexor and is involved in pronation and supination.)

ii. extensor carpi radialis longus and brevis—to the metacarpals

iii. extensor digitorum (communis)—to the middle and distal phalanges

iv. extensor digiti minimi—to the extensor digitorum communis expansion (proximal phalanx 5th digit)

v. extensor carpi ulnaris—to the 5th metacarpal

➢ 2. *Deep group (415)—arise from the dorsal aspect of the bones of the forearm and act on the thumb and index finger.*
The deep group is innervated by the radial nerve.

i. supinator—from the lateral epicondyle to the radial shaft

ii. extensor pollicis longus and brevis—from the interosseous membrane to the phalanges

iii. abductor pollicis longus—from the radius to the thumb metacarpal

iv. extensor indicis (proprius)—from the radius to the extensor tendon of the index finger

V. EXTENSORS IN THE POSTERIOR FOREARM ACCORDING TO ACTION (411)

A. Act on the elbow joint—anconeus, supinator, and brachioradialis

B. Extensors of the wrist—extensor carpi ulnaris, extensor carpi radialis and brevis

C. On the thumb—abductor pollicis longus, extensor pollicis longus and brevis

D. Extensors of MP joints—extensor digitorum communis, extensor digitorum indicis, and extensor digiti minimi

VI. MUSCLES WHICH MOVE THE FOREARM (402, 403)—in addition to the brachioradialis, include the brachialis and triceps

VII. BRACHIAL ARTERY (405)—ends near the neck of the radius to divide into the radial and ulnar arteries. The median nerve accompanies the brachial artery into the forearm.

VIII. ULNAR ARTERY (418)

A. Course—The ulnar artery begins below the bend of the elbow and goes down to the radial side of the pisiform bone. Toward the wrist, it becomes more superficial.

B. Branches

1. *Recurrent—to forearm muscles*
2. *Muscular—to ulnar side muscles*
3. *Interosseous arteries*
4. *Carpal—to carpals and* ***anastomoses*** *with radial branches*
5. *Volar (Anterior) arch—curves across the palm convex downward and has branches to the volar digital arteries*

IX. RADIAL ARTERY (418, 435)

A. Course—The radial artery begins at the bifurcation of the brachial below the elbow bend and continues along the radial side of the forearm, around the lateral carpals, and into the palm to unite with the volar (a branch of the ulnar) to form the deep volar arch.

B. Branches—supply the radial portions of the forearm, wrist, and hand

1. *Muscular branches—on the radial sides of the forearm and hand*
2. *Branches to carpals and metacarpals*
3. *Radial recurrent—joins a branch of the deep brachial artery*
4. *Palmar carpal—leaves above the wrist and joins the ulnar carpal*
5. *Superficial volar—thru the short muscles of the thumb to join the ulnar and form the deep volar arch*

X. VEINS (449)—(have valves)

A. Superficial Veins—larger than the deep veins with which they communicate.

1. *Basilic—extends from the back of the hand to the brachial vein*
2. *Cephalic—from the radial end of the dorsal venous arch of the hand to the axillary vein*
3. *Median cubital—In the antecubital fossa, it connects the cephalic and basilic veins.*

B. Deep Veins—accompany arteries—begin in the hand as volar digital veins and end by draining into the axillary vein

XI. NERVES (444, 445, 447)

A. Sensory

1. *Median—to the articular surface of the elbow joint*
2. *Ulnar—to the articular surface of the elbow joint*
   i. dorsal cutaneous branch
3. *Radial (posterior interosseous branch)—a superficial branch which supplies the skin of the forearm*

B. Motor

1. *Median (418, 419)—related to pronator teres and flexor digitorum profundus. It supplies the pronators and most of the flexors and lies between the superficial and deep flexors of the fingers.*
2. *Ulnar (418, 419)—supplies the flexor carpi ulnaris and flexor digitorum profundus*
3. *Deep branch of radial (415)—supplies the supinator and posterior fascial compartment and the extensors of the thumb*

Wrist

I. WRIST BONES (422, 423, 425)

A. Radius and Ulna

B. Carpals (from the radial to the ulnar side)

1. *Proximal row*

i. scaphoid (navicular)—the only bone to traverse the proximal and distal carpal rows. In addition to the radius, it articulates with the lunate, capitate, and trapezium.

ii. lunate

iii. triquetrum

iv. pisiform—below the distal flexor wrist crease on the ulnar side

II. WRIST JOINT (423, 425)—The scaphoid, lunate, and triquetral articulate with the radius and articular disk of the ulna.

A synovial cyst arising from the dorsal radiocarpal joint, or a bursa of a tendon sheath, is called a "ganglion."

III. WRIST INJURIES

A. **A fall on the extended hand may:**

1. ***Result in a fracture of the scaphoid (navicular) bone***

2. ***Dislocate the lunate and affect the tendons of the flexors of the digits or injure the median nerve***

B. **The ulnar artery may be injured in superficial lacerations of the wrist.**

IV. CAPSULE (425)

A. Radiocarpal Ligaments—from the styloid process of the radius to the carpal bones and ligaments

B. Ulnar Collateral Ligament—from all sides of the styloid process of the ulna to the pisiform and triquetral

C. Radial Collateral Ligament—from the styloid process of the radius to the scaphoid and capitate

V. CARPAL TUNNEL (430)—an osteofascial tunnel for the flexor tendons and median nerve

A. Flexor Tendons—pollicis longus, digitorum profundus, and digitorum superficialis

B. Median Nerve—passes deep to the retinaculum

C. **Tests for Carpal Tunnel Syndrome (CTS) (compression neuropathy of the median nerve resulting in sensory changes and muscular weakness)**

1. ***For paresthesia***

   i. **fingertip sensitivity—of index, middle, and ring fingers and thumb**

   ii. **Phalen's maneuver—wrist flexion**

   iii. **Tinel's nerve percussion—wrist tap above the median nerve**

2. ***For atrophy***

   i. **strength test—a flattening of the flexor pollicis longus**

➢ VI. SYNOVIAL SHEATHS (430, 439)

➢ A. Anterior Sheaths (430)—The median nerve lies between the radial and ulnar bursae.

1. *Ulnar bursa—for flexor tendons (digitorum sublimis and profundus)*
2. *Radial bursa—for the flexor pollicis longus*
3. *Sheath for the flexor carpi radialis*

➢ B. Posterior (Extensor) Sheaths (439)—lie under the extensor retinaculum. Extensor sheaths include those for: carpi ulnaris, digiti minimi, extensor digitorum and indicis, pollicis longus, carpi radialis longus, carpi radialis brevis, and pollicis brevis.

A stenosing tenosynovitis of the tendon sheaths of the extensor pollicis brevis and the abductor pollicis longus is called "de Quervain's disease."

C. Abductor Sheath—for the pollicis longus

➢ VII. RETINACULA (424)—keep tendons close to the wrist bones

➢ A. Extensor Retinaculum (439)—1 inch wide—runs from the radial styloid over the ulnar styloid to the deep fascia in the medial palm—The extensor tendons to the fingers pass beneath it.

➢ B. Flexor Retinaculum (424, 430)—stretches between carpal bones (hamate, pisiform, scaphoid, and trapezium) to form the roof of the carpal tunnel

1. *Volar carpal ligament—over the ulnar artery and nerve*
2. *Transverse carpal ligament—the roof of the carpal tunnel—over the median nerve and 3 long flexors*

VIII. ARTERIES (430)—Articular arteries arise from the carpal arterial arches of the ulnar and radial arteries.

A. Radial Artery in the wrist (424)—lies superficial to the distal end of the radius and lateral to the flexor carpi radialis tendon

B. Ulnar Artery in the wrist (424)—lies superficial to the flexor retinaculum and lateral to the ulnar nerve

IX. NERVES PASSING THRU THE WRIST (416–418, 429, 430)

A. Ulnar—lies superficial to the flexor retinaculum and medial to the ulnar artery

Ulnar nerve injury at the wrist produces sensory changes to the hand and affects the interosseous muscles.

B. Median—lies medial to the flexor carpi radialis tendon

C. Radial—**Radial nerve injury may cause "wrist drop."**

1. *Superficial branch—to the lateral portion of the dorsum*

2. *Deep branch—supplies the lateral surface of the hand and lateral digits to the proximal IP joints*

Hand

I. BONES (426)

A. Carpals (Wrist bones) (425, 426)

1. *Proximal—scaphoid, lunate, triquetral, and pisiform. (The pisiform is below the distal flexor wrist crease on the ulnar side.)*
2. *Distal—trapezium, trapezoid, capitate, and hamate*

B. Metacarpals (426)

C. Phalanges (426)

1. *Thumb—proximal and distal*
2. *Fingers—proximal, middle, and distal*

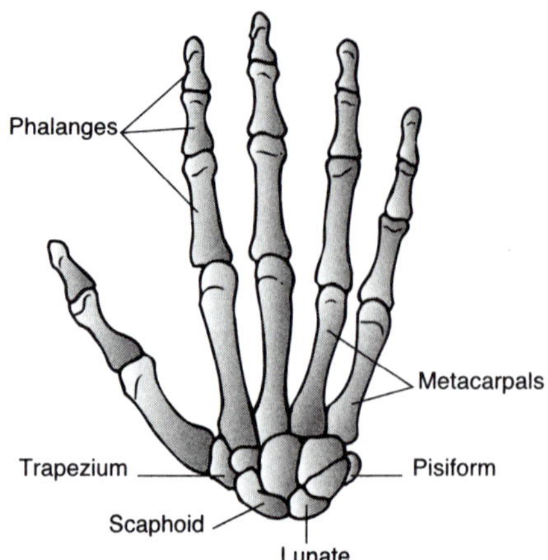

II. FRACTURES

A. **Metacarpals—due to an unauthorized fistfight**

B. **Phalanges—The direction of bone displacement depends on whether the fracture occurs proximal or distal to a particular tendon insertion.**

III. FINGER MUSCLES (412, 413, 432–434)

A. Extrinsic

1. *Extensors—innervated by branches of the* ***radial*** *nerve—1 extensor tendon to the middle finger and 1 to the ring finger—2 extensor tendons to the index and little fingers. The extensor tendons straighten the fingers.*

i. arise from a common attachment on the lateral epicondyle of the humerus to extend thru the wrist in extensor sheaths and held in place by the extensor retinaculum. Extensor tendons insert on the extensor expansion (hood).

a. extensor carpi radialis brevis—from the humerus to the metacarpals—extends the hand

b. extensor digitorum—from the lateral epicondyle of the humerus to the extensor expansion of the middle and distal phalanges—extends the fingers

c. extensor digiti minimi—from the humerus to the extensor expansion of the proximal phalanx of the 5th digit—extends the little finger

d. extensor carpi ulnaris—from the lateral epicondyle of the humerus to the 5th metacarpal—extends the hand

ii. arise from the supracondylar ridge of the humerus

a. extensor carpi radialis longus—from the humerus to the metacarpals—extends and abducts the hand

➢ 2. *Flexors* (412, 413)

i. innervated by the **ulnar** nerve

a. flexor carpi ulnaris—from the ulna and medial epicondyle of the humerus to the hamate and 5th metacarpal

b. flexor carpi profundus

ii. innervated by the **median** nerve

a. flexor digitorum superficialis—from the ulna and humerus to the middle phalanges. It inserts on the middle phalanges and flexes the proximal IP joint. Its tendon splits to allow for passage of the profundus.

The function of the flexor digitorum superficialis can be tested by holding the fingers in extension while the subject attempts to flex 1 finger. This removes the effect of the profundus.

A stenosing tenosynovitis of the flexor tendons passing thru the "pully mechanism" may produce a "trigger finger."

b. flexor digitorum profundus—from the ulna to the distal phalanges—the only muscle which flexes the distal IP joint

c. flexor carpi radialis—from the humerus to the 2nd and 3rd metacarpals

➢ B. Intrinsic (434) (They position the digits.)

1. *Interossei—extend from the metacarpals to the phalanges—innervated by the **ulnar nerve***

i. dorsal interossei (4)—from **adjacent** metacarpals to the proximal phalanges (extensor expansion)—flex, extend, and **abduct** the fingers

ii. palmar interossei (3)—from the metacarpals **across MP** joints to the proximal phalanges (extensor expansion)—flex, extend, and **adduct** the fingers

➢ 2. *Lumbricals* (432)*—innervated by the **median nerve**—from the tendon of the flexor digitorum profundus to the terminal phalanges (extensor expansions) and capsules of MP articulations. The lumbricals flex the fingers at MP joints and extend fingers at IP joints.*

IV. THUMB MUSCLES (429, 434)

A. Extrinsic

1. *Extensors—innervated by branches of the* ***radial*** *nerve*

   i. extensor pollicis longus and brevis—from the interosseous membrane to the phalanges of the thumb (the longus to the distal phalanges and the brevis to the proximal phalanx)

   ii. abductor pollicis longus—from the interosseous membrane to the metacarpals of the thumb—produces radial deviation of the wrist (abducts thumb and wrist)

2. *Flexors—innervated by the* ***median*** *nerve*

   i. flexor pollicis longus—from the interosseous membrane and radius to the distal phalanx of the thumb. Its tendon bridges all thumb joints.

B. Intrinsic—innervated by the recurrent branch of the **median** nerve

1. *Short muscles of the thenar eminence—abductor pollicis brevis, flexor pollicis brevis, opponens pollicis, and adductor pollicis. They are involved in abduction, pronation, flexion, and opposition of the thumb. They cup the palm and spread the hand.*

   Thumb opposition can be used as a test of median nerve function in the hand or a test of opponens pollicis function.

C. Intrinsic Thumb Muscles—innervated by the **ulnar** nerve

1. *Thenar muscles—adductor pollicis*

2. *Hypothenar muscles—abductor, flexor, and opponens digiti minimi—from the carpal bones or flexor retinaculum—abduct or flex the fingers*

V. FASCIA OF THE BODY OF THE HAND (428)

A. Dorsal Fascia

1. *Superficial*

   i. blends with the extensor retinaculum

   ii. The dorsal veins and nerves are located superficial to the superficial dorsal fascia.

   iii. forms 2 superficial spaces

      a. subcutaneous

      b. subaponeurotic

2. *Deep dorsal fascia covers the interossei and adductors of the thumb.*

B. Palmar Fascia (428, 429)

1. *Palmar (Deep) Aponeurosis—a latticework of fascia which compartmentalizes tendons, lumbrical muscles, nerves, and the muscles which extend the fingers. It extends from the extensor retinaculum to the digital fascial sheaths. The palmar aponeurosis lies deep to the superficial fascia and above the tendon of the long flexors, lumbricals, and interossei.*

Palmar and digital contraction due to fasciitis of the palmar aponeurosis is called "Dupuytren's Contracture."

2. *Deep fascia—located below the palmar aponeurosis and overlies the volar interossei and the thumb adductor*

VI. SPACES OF THE BODY OF THE HAND (429, 431)

A. Dorsal Spaces (The digital extensor tendons are located between 1. and 2.)

1. *Dorsal subcutaneous space—located above the extensor tendons*
2. *Dorsal subaponeurotic space—located below the extensor tendons*

B. Palmar spaces

1. *Midpalmar space—located deep to the ulnar bursa, lumbrical muscles, flexor tendons, and palmar digital arteries and nerves*
2. *Thenar space—located deep to the flexor indicis tendon and the lumbrical muscle*
3. *Web spaces—a thinning of the fascia between the dorsum and palm of the hand which may allow infection to spread from the palm to the dorsum. (The cutaneous web space of the thumb is supplied only by the radial nerve.)*

VII. FASCIA OF THE FINGERS AND THUMB (429)

A. Dorsal Fascia

1. *Fascia of the distal dorsal phalanges—extends from the nail to the joint and bone*
2. *Retinacular bands*

B. Volar Septae—connect the volar fat pad with the bone of the flexor sheath

VIII. SYNOVIAL BURSAE (429, 431, 432)

A. Radial—from the distal forearm to the thumb. The radial bursa may communicate with the ulnar bursa.

B. Ulnar of the Little Finger—extends from the lower forearm thru the carpal tunnel and surrounds the long flexors of the wrist

C. Bursae around the Tendons of the three middle fingers extend to the midportion of the palm.

IX. JOINT MOVEMENTS

A. Carpometacarpal Joint Movements

1. *Thumb*

   i. Opposition involves flexion, medial rotation, and abduction. The thumb can touch the fingers.

   ii. Reposition involves abductor and extensor pollicis longus muscles. The thumb is returned to the anatomic position.

2. *Fingers—The 4 metacarpals articulate with the distal row of carpals. The fingers have strong collateral ligaments.*

   i. intrinsic—related to the middle (long) finger. (The movement is due to the interossei and lumbricals.)

      a. abduction—away from the middle finger

      b. adduction—toward the middle finger

   ii. extrinsic (The movement is due to extensors and flexors.)

B. Metacarpophalangeal Movements—Metacarpals articulate with the proximal phalangeal bones. Abduction, adduction, and ulnar and radial deviation of the hand occur at the MP joints.

Dislocations may occur at these ball-and-socket joints.

1. *Thumb—opposition—involves flexion at MP joint*

C. Interphalangeal (IP) Movements—between phalanges

1. *Involve flexion and extension*

2. *Have strong collateral ligaments*

3. *The IP joints are hinge joints.*

D. Intercarpal Movements

X. MOVEMENTS OF THE HAND AND FINGERS

A. Hand Movements

1. *Ulnar deviation—angulation of the hand to the ulnar side—a function of the flexor and extensor carpi ulnaris*

2. *Radial deviation—angulation of the hand toward the radial side—a function of the abductor pollicis longus*

3. *Pronation—movement of the palm to face down*

4. *Supination—the palm is up*

B. Finger Movements

1. *Abduction—movement of the fingers from the midline of the long finger—a function of the dorsal interossei*

2. *Adduction—movement of the fingers toward the midline of the long finger*

➢ XI. ARTERIES (435, 438, 440)

A. Radial Artery

1. *Superficial palmar—contributes to the palmar arch*

2. *Dorsal carpal*

   i. dorsal digitals

   ii. dorsal pollicis—to the dorsal aspects of the thumb

3. *Palmar arteries—to thumb and index finger*

4. *Deep palmar arch*

B. Ulnar Artery (may be compressed over the hypothenar eminence just lateral to the pisiform bone)

1. *Dorsal carpal—to the dorsal aspects of the hand*

   i. dorsal metacarpals

      a. dorsal digital arteries

2. *Deep palmar artery*

3. *Superficial palmar arch*

   i. palmar digital arteries—to the fingers

➢ XII. PALMAR ARTERIAL ARCHES (435)

A. Superficial (Volar)—a continuation of the ulnar artery. It partially lies between the transverse creases. A small superficial palmar artery (branch of the radial) contributes to the superficial palmar arch.

B. Deep—a continuation of the radial artery. The deep arch lies proximal to the superficial arch. It receives blood from the radial and ulnar arteries.

XIII. VEINS (437)—drain into cephalic and basilic veins

A. Dorsal Veins (prominent on the dorsum of the hand)

1. *Digital veins*

B. Palmar Veins

1. *Digital veins*

XIV. SUPERFICIAL VENOUS ARCH (437)—Flow is from the palmar to the dorsal side of the hand.

A. Medial portion—flows into basilic vein

B. Lateral portion—flows into cephalic vein

XV. LYMPHATICS (452)—generally follow the course of veins, initially to the dorsal subcutaneous (fascial) spaces and thence to drain into the epitrochlear (supratrochlear) and axillary nodes

XVI. NERVES (434–438, 441, 444, 445)—The median, radial, and ulnar nerves supply the hand.

A. Sensory

1. *Superficial branch of the radial—supplies the posterior surface of the hand and lateral 2 digits*

2. *Ulnar—supplies the surface of the medial part of the back of the hand. The ulnar nerve is the only sensory supply to the little finger.*

3. *Palmar cutaneous branch of the median—supplies the skin of the central-lateral portion of the palm and the thumb and 2½ fingers*

B. Motor (no motor nerves on the dorsum of the hand)

1. *Ulnar—supplies the flexor carpi ulnaris, flexor digitorum profundus, hypothenar muscles, medial lumbricals, and adductor pollicis.*

   Ulnar nerve injury may cause "claw hand."

2. *Median—supplies 1st and 2nd lumbricals and 3 thenar muscles (abductor and flexor pollicis brevis and opponens pollicis).*

   Median nerve injury may cause inability to flex and pronate the thenar muscles and fully flex the index and middle fingers.

Lower Limb

Page numbers in black. Netter plate numbers in green.

Lower Limb—General

I. MUSCLES (458–461, 481–486, 497–499)

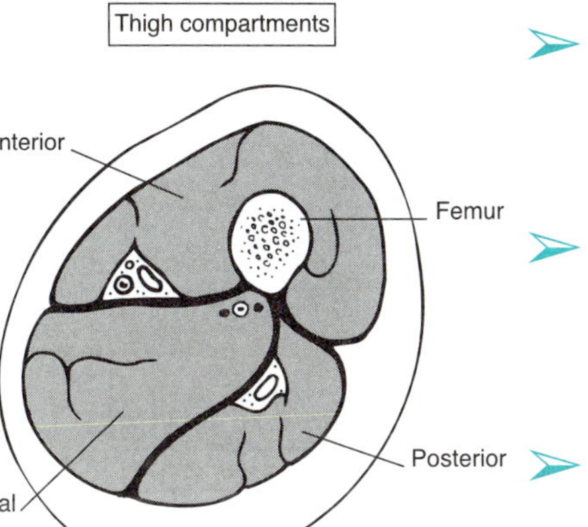

A. In the Thigh (458–461)

1. *Anterior group (thigh flexors) (458, 459)—separated from the medial and hamstring groups by septa—consist of the sartorius, quadriceps femoris, pectineus, and iliopsoas—innervated by the* ***femoral*** *nerve (L2–L3)*
2. *Medial group (adductors) (459)—gracilis and obturator externus—innervated by the* ***obturator*** *nerve (L2–L3)*

   A "groin pull" affects the anterior and medial thigh group muscles.

3. *Hamstring group (knee flexion and thigh extension) (461)—long head of the biceps femoris, semitendinosus, and semimembranosus—innervated by the* ***branches of the sciatic nerve***
4. *Gluteal (thigh abductor group) (classified as gluteal muscles) (461)—gluteus medius and minimus (tensor fascia lata)—innervated by the* ***superior gluteal*** *nerve (L4-S1)*
5. *Gluteal (thigh extensor muscle) (460–461)—gluteus maximus—innervated by the* ***inferior gluteal*** *nerve (L5-S2)*

B. In the Leg (481–486)—innervated by branches of the **sciatic** nerve

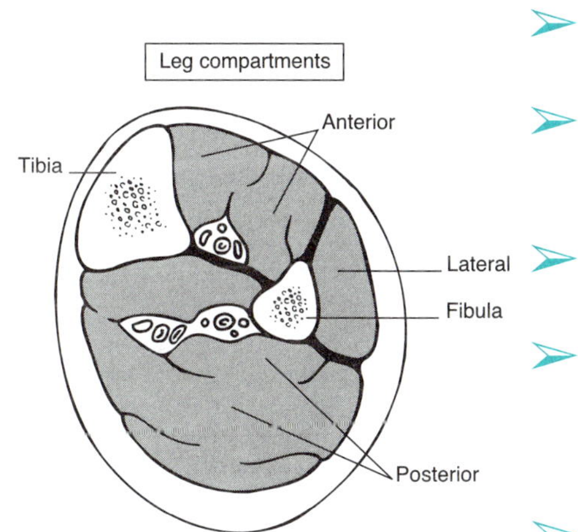

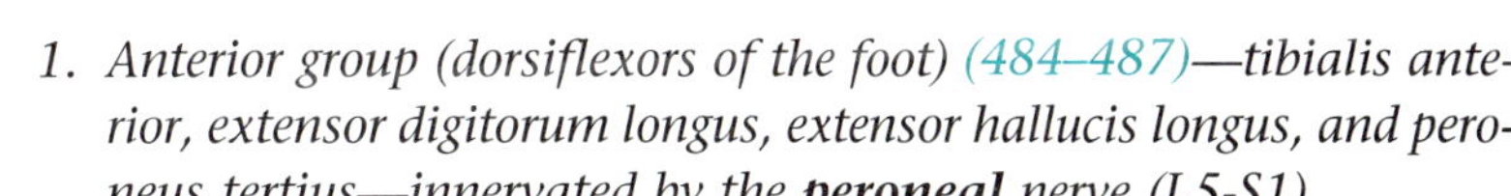

1. *Anterior group (dorsiflexors of the foot) (484–487)—tibialis anterior, extensor digitorum longus, extensor hallucis longus, and peroneus tertius—innervated by the* ***peroneal*** *nerve (L5-S1)*
2. *Lateral group (flex the ankle and evert the foot) (486)—peroneus longus and brevis—innervated by the* ***peroneal*** *nerve (L5-S2)*
3. *Posterior group (481–483)—soleus, gastrocnemius, and plantaris (triceps surae) (plantar-flexors of the foot), and other deep muscles of the leg—innervated by branches of the* ***tibial****. The Achilles tendon (from the triceps surae) inserts on the calcaneus.*

C. Intrinsic Muscles in the Foot (495, 497–499)

1. *The short extensors are innervated by the* ***peroneal*** *nerve.*
2. *The plantar muscles are innervated by the plantar branches of the* ***tibial*** *nerve.*

II. FEMORAL ARTERY (477)—After passing below the inguinal ligament, the external iliac artery becomes the femoral. It gives off branches to the lower anterior abdominal wall (superficial epigastric and superficial circumflex iliac).

Injuries to the femoral or popliteal arteries may result in an aneurysm.

A. The main branch of the femoral in the thigh is the **profunda femoris** which proceeds downward posteriorly to give off nutrient branches and perforating branches (capable of supplying a major portion of the needs of the lower limb if the femoral is obstructed). In the thigh, the femoral has additional branches including the inguinal, circumflex, and upper genicular.

B. Just above the knee and after it passes thru the hiatus in the adductor magnus, the femoral becomes the **popliteal**. The sural branch supplies calf muscles, and genicular branches go to the knee. The popliteal divides into the anterior and posterior tibials.

1. *The anterior tibial continues down to the foot to become the dorsalis pedis.*
2. *The posterior tibial forms **anastomoses** with the anterior tibial to supply the toes.*

III. FEMORAL VEIN (508, 509)—Its tributaries have valves.

A. Superficial Veins

1. *The great (greater) (long) saphenous begins in the dorsum of the foot along the medial border of the tibia (behind the medial condyle of the tibia and behind the medial condyle of the femur) and extends upward to enter the femoral vein.*
2. *The small (lesser) (short) saphenous begins in the foot (posterior to the lateral malleolus) and extends along the back of the calf to enter the popliteal fossa and vein.*

B. Deep Veins—The anterior and posterior popliteals drain into the popliteal which becomes the femoral.

IV. INGUINAL LYMPH NODES (510)

A. Superficial—A horizontal group drains the perineum and part of the lower abdominal wall. A vertical group follows the course of the saphenous veins.

B. Deep—located near the upper femoral artery. Their afferents follow the course of the femoral vein. The deep nodes also receive afferents from the popliteal nodes.

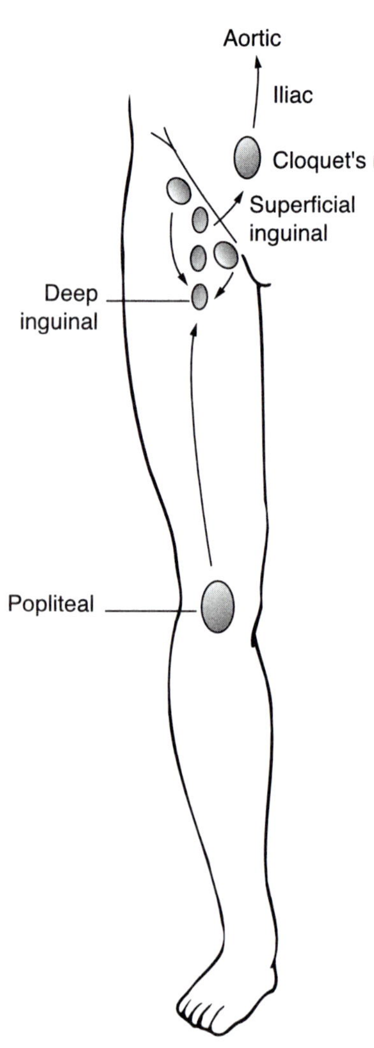

V. NERVES (463–465)

A. Lumbar Plexus (464)—Branches in the thigh:

1. *Obturator—sensory from the skin over a portion of the medial thigh and motor to the medial adductor group*
2. *Femoral branch of the genitofemoral—sensory from the small area over the femoral triangle*
3. *Lateral femoral cutaneous—sensory from the anterolateral thigh*
4. *Ilioinguinal—sensory to the medial thigh*
5. *Femoral—sensory fibers from the anterior and medial thigh and motor to the anterior thigh muscles (sartorius, pectineus, and quadriceps)*

   Injury may affect flexion at the knee and cause a decrease in the patellar reflex.

B. Sacral Plexus (463)

1. *Gluteal nerves (via the greater sciatic foramen)*
   - i. superior—to the thigh abductor group
   - ii. inferior—to the gluteus maximus (a thigh extensor)
2. *Nerves to the external thigh rotators (piriformis, obturator internus, gemelli, and quadratus femoris)*
3. *Sciatic nerve—supplies a good portion of the lower limb*
   - i. branches to the posterior thigh muscles

     May be injured in a posterior hip dislocation
   - ii. posterior femoral cutaneous nerve of the thigh
   - iii. tibial (medial popliteal)—to the posterior leg and plantar foot muscles

     Injury may impair the ability to stand on, abduct, or adduct the toes.
   - iv. common peroneal (common fibular) (lateral popliteal)—sensory to the anterior leg and foot and motor to the anterior group of leg muscles and the short extensors of the foot

     The peroneal nerve is usually injured as it winds around the neck of the fibula. The dorsiflexors and evertors of the foot may be affected.
4. *Sural (504–506)—formed by the tibial and common peroneal—sensory from the lateral leg and foot*

Arteries to the Lower Limb

I. OBTURATOR (470)—a branch of the internal iliac. Branches in the lower limb:

A. Muscular Branches

B. Artery in the Ligament (ligamentum teres) to the Head of the Femur (usually insufficient to maintain a blood supply to the femoral head by itself)

II. FEMORAL (466, 467, 477)—supplies the lower limb and sends branches to the anterior lower abdominal wall. The femoral artery is a continuation of the external iliac.

A. Course—The continuation of the external iliac begins behind the inguinal ligament in the femoral sheath. The nerve lies laterally and the vein medially. The lower portion of the femoral lies medial to the femur. The femoral artery ends at the junction of the middle ⅓ and lower ⅓ of the thigh where it passes thru an opening in the adductor magnus, called the adductor or subsartorial canal, to become the popliteal artery.

B. Branches

1. *Superficial epigastric—arises from the front of the femoral and turns upward to ascend in the superficial fascia of the anterior abdominal wall to the umbilicus*

2. *Superficial circumflex iliac—runs in the direction of the inguinal ligament to supply the skin*

3. *Superficial external pudendal—extends medially to the skin of the lower abdomen*

4. *Deep external pudendal—extends medially to the skin of the scrotum or labium majus*

5. *Muscular branches—to the sartorius, vastus medialis, and adductors*

C. Profunda Femoris (Deep Femoral) branch—supplies the anterior compartment and contributes to hip and knee joints

1. *Course—arises posterolaterally and continues behind the femoral vein*

2. *Branches*

i. medial femoral circumflex—to adductors, obturator externus, and acetabulum

ii. lateral femoral circumflex—extends around the femur below the greater trochanter to the rectus femoris and vastus lateralis muscles. (**NOTE:** Both medial and lateral circumflex arteries form a ring around the neck of the femur and penetrate the capsule of the hip joint to supply the head of the femur.)

iii. perforating—Branches perforate the adductor magnus and go to the back of the thigh. The perforating branches extend from the anterior to the posterior compartment and allow for collateral circulation to the lower limb when the femoral artery is compromised (below the level of the origin of the profunda).

D. Highest Genicular—to the knee joint

III. POPLITEAL (483)—supplies the hamstrings and knee joint

A. Course—A continuation of the femoral, it extends downward behind the lower femur. In the popliteal fossa, the popliteal artery lies below the vein and tibial nerve. It ends as the anterior and posterior tibial arteries.

B. Branches

1. *Muscular—to adductor magnus and hamstring muscles*
2. *Sural—to gastrocnemius, soleus, and plantaris muscles*
3. *Genicular—to the knee*

IV. ANTERIOR TIBIAL (477, 485, 494, 495)—supplies the extensor compartment of the leg

A. Course—begins at the lower border of the popliteus and passes down the lateral side of the tibia, deep to the inferior extensor retinaculum, to end as the dorsalis pedis at the front of the ankle

B. Branches

1. *To the muscles of the anterior crural compartment*
2. *Tibial recurrent arteries which join the* ***anastomoses*** *around the knee*
3. *Malleolar branches which supply the lateral malleolus and* ***anastomoses*** *around the ankle*

➢ **V. DORSALIS PEDIS (494, 495)—to the anterior part of the foot. It is the continuation of the anterior tibial artery.**

A. Course—along the tibial side of the dorsum of the foot in front of the articular capsule of the ankle joint

The dorsalis pedis artery can be palpated lateral to the extensor hallucis longus tendon.

B. Branches

1. *Deep plantar—lies deep to the flexor retinaculum to go to the great toe and contribute to the formation of the plantar arch*
2. *Tarsals—to tarsals and muscles*
3. *Arcuate—to metatarsals*

➢ **VI. POSTERIOR TIBIAL (483)—supplies the posterior crural compartment**

A. Course—The posterior tibial artery descends along the tibia in the posterior compartment, to pass around the medial malleolus to the sole of the foot and divide into plantar branches.

B. Branches

1. *Fibular (Peroneal) artery—extends along the fibula*
   i. supplies muscles of the posterior crural compartment
   ii. supplies muscles of the lateral crural compartment
   iii. is the nutrient artery to the fibula
   iv. supplies the ankle joint
   v. **anastomoses** with the arcuate artery on the dorsum of the foot
2. *Circumflex fibular artery—contributes to the **anastomoses** around the knee*
3. *Nutrient artery—to the tibia*
4. *Calcanean arteries—supply the heel*
5. *Malleolar branch—to the medial malleolus*
6. *The posterior tibial artery terminates beneath the flexor retinaculum by dividing into plantar arteries.*
   i. medial plantar—to the medial side of the foot
   ii. lateral plantar—to the lateral side of the foot. It joins the deep plantar branch of the dorsalis pedis in forming the plantar arch.

Nerves to the Lower Limb

I. LUMBAR PLEXUS (462–464)

A. Obturator (L2–L4)—supplies the medial thigh (motor to the adductor longus, adductor brevis, and gracilis) and has articular branches to the hip and knee joints

Injury may result in impaired adduction of the thigh.

B. Femoral branch of the Genitofemoral (L1–L2)—to the skin over the femoral triangle

C. Lateral Cutaneous Nerve of the Thigh (L2–L3)—to the anterolateral thigh

D. Ilioinguinal (L1)—to the anterior scrotum and medial thigh

II. FEMORAL (466, 467, 502)—supplies the muscles of the anterior thigh and the hip and knee joints and has sensory branches along the medial aspect of the lower limb

A. In the Thigh

1. *Motor—to the iliacus, sartorius, pectineus, and quadriceps*
2. *Sensory—intermediate and medial cutaneous nerves of the thigh, except posteriorly*

B. Saphenous Nerve—the terminal branch of the femoral

1. *In the leg—sensory to the skin of the front and medial side*
2. *In the foot—sensory to the posterior half of the dorsum and medial aspect of the foot*

III. SCIATIC (L4-S3) (463, 468, 469, 504)—enters the gluteal region thru the greater sciatic foramen, deep to the gluteus maximus, then enters the thigh between the ischial tuberosity and the greater trochanter of the femur. It runs down the back of the thigh on the posterior surface of the adductor magnus. In the superior portion of the popliteal space, the sciatic nerve divides into the Tibial (Medial Popliteal branch) and the Common Peroneal (Lateral Popliteal branch).

A. In the Thigh—posterior cutaneous nerve of the thigh

IV. TIBIAL (MEDIAL POPLITEAL) (L4-S3) (468, 505)—supplies the hamstring muscles

A. In the Knee—It passes thru the popliteal space and then extends down the leg under the gastrocnemius and soleus.

B. In the Upper Leg—The tibial nerve lies under the soleus and gastrocnemius (hamstrings) to supply the deep muscles of the dorsum of the leg and the fibula.

Injury to the tibial nerve results in inability to stand on the toes or adduct the toes.

C. In the Ankle—The tibial nerve lies under the calcaneomalleolar ligament (i.e., passing between the medial malleolus and the calcaneous) and innervates the ankle joint.

D. In the Foot—The tibial nerve innervates the heel.

1. *Medial plantar nerve*
   i. motor to the abductor hallucis, flexor digitorum brevis, medial lumbrical, and flexor hallucis brevis
   ii. sensory to the medial side of the foot
2. *Lateral plantar nerve*
   i. motor to the interossei, abductor digiti minimi, quadratus plantae, lateral 3 lumbricals, flexor digiti minimi brevis, and adductor hallucis
   ii. sensory to the lateral side

➢ **V. COMMON PERONEAL (COMMON FIBULAR) (LATERAL POPLITEAL) (L4-S2)** (506)**—continues down a short distance along the posterior border of the biceps femoris and behind the soleus. At the neck of the fibula it curves around its lateral side,** where it is susceptible to injury. **It then divides into superficial and deep branches.**

A. Recurrent Articular—supplies the knee joint

➢ B. Deep Peroneal (Deep Fibular) (Anterior tibial) (485)—a continuation of the common peroneal. It extends forward between the extensor digitorum and the tibialis anterior into the anterior compartment of the leg.

Injury may result in "foot drop."

1. *In the leg—passes anterior to the interosseous membrane*
   i. motor—to the anterior leg muscles (tibialis anterior, extensor hallucis longus, and extensor digitorum longus)
   ii. sensory—lateral cutaneous nerve to the calf of the leg
2. *In the ankle—to the ankle joint*
3. *In the foot*
   i. medial digital branch to the skin of the lateral side of the big toe
   ii. motor—to the extensor digitorum brevis

➢ C. Superficial Peroneal (Musculocutaneous) (L5-S2) (485)—supplies the lateral compartment of the leg

Injury may result in impaired eversion of the foot.

1. *In the leg—pierces the peroneus longus and brevis and descends superficially in the lower portion of the leg*

    i. motor to the peroneus longus and brevis

    ii. sensory dorsal cutaneous branches to the anterolateral leg

➢ **VI. SURAL (CUTANEOUS) (505, 506)—formed in the popliteal fossa from sensory branches of the tibial and deep peroneal (deep fibular) nerves. It descends along the Achilles tendon with branches to the inferolateral part of the leg.**

A. In the Leg—lateral sural cutaneous (S1–S2)—innervates the skin distal to the midcalf (lateral side of the lower leg)

B. In the Ankle—descends along the lateral border of the Achilles tendon to pass dorsal to the lateral malleolus

C. In the Foot—The lateral sural (dorsal) cutaneous (S1–S2) innervates the skin of the lateral aspect of the foot.

Hip

I. BONES (453, 454)

A. "Hip Bone"—formed by the ilium, ischium, and pubic bones which meet in the acetabulum

B. Femur

1. *Head—covered with cartilage*

   i. fits into the acetabulum—a synovial ball-and-socket joint

   Fracture of the acetabulum may result in degenerative arthritis.

   ii. contains the *fovea* for the attachment of the ligamentum teres

   iii. glenoid labrum—a cartilagenous ring which surrounds the head of the femur and helps hold it in place

2. *Neck—connects the head and body*
3. *Greater trochanter—attachment for the gluteals, gamelli, piriformis, and obturator internus muscles*
4. *Intertrochanteric (Spiral) line—attachment for the iliofemoral ligament*
5. *Lesser trochanter—attachment for the psoas major muscle*

II. LIGAMENTS (454)

A. Ligaments around the Neck of the Femur

1. *Iliofemoral—located laterally and extends from the iliac bone to the intertrochanteric line*
2. *Ischiofemoral—spirals from the iliac bone posteriorly to the intertrochanteric line of the femur posteriorly*
3. *Pubofemoral—from the inferior part of the acetabular rim to the femur*

B. Ligament to the Head of the Femur (Ligamentum Teres)—extends from the acetabular notch to the *fovea* and contains the "artery in the ligament to the head of the femur"

C. Transverse Ligament—over the acetabular notch

III. MUSCLES (461, 468, 469)

A. Muscles Which Abduct the Thigh (468, 469)

1. *And extend from the ilium to the greater trochanter—gluteus medius and minimus*
2. *And insert on the iliotibial tract—tensor fascia lata*

B. Muscles Which Externally Rotate the Thigh (469)

1. *And insert on the greater trochanter*

i. piriformis—from the sacrum

ii. obturator internus—from the obturator membrane

iii. superior gemellus—from the ischial spine

iv. inferior gemellus—from the ischial tuberosity

2. *And insert on the quadrate tubercle of the femur—quadratus femoris*

C. Muscles Which Extend the Thigh (461)

1. *And insert on the iliotibial tract—gluteus maximus*

IV. MOVEMENTS AT THE HIP JOINT

A. Flexion—iliopsoas, rectus femoris, tensor fasciae latae, pectineus, sartorius

B. Extension—long head of biceps femoris, semitendinosus, semimembranosus, gluteus maximus

C. Adduction—adductor magnus, adductor longus, adductor brevis, gracilis, obturator externus

D. Abduction—gluteus medius, gluteus minimus, sartorius

E. Medial rotators—gluteus medius, gluteus minimus

F. Lateral Rotation—obturator internus, obturator externus, quadratus femoris, and piriformis (when the hip is extended)

V. ARTERIES (468, 470)

A. Medial and Lateral Circumflex Arteries (from the profunda femoris)—form a ring around the neck of the femur and penetrate the capsule of the hip joint and supply the head of the femur

B. Artery in the Ligament to the head of the femur

C. Gluteal Arteries (from the internal iliac)—to the acetabulum and gluteal muscles

VI. LYMPHATICS—to the obturator and iliac nodes

VII. NERVES (469)

A. Articular Nerves—from the femoral and obturator nerves

B. Sensory—posterior cutaneous nerve of the thigh

C. Motor—The sacral plexus innervates the piriformis, obturator internus, gemelli, and quadratus femoris.

Thigh

I. FEMUR BONE (455)—longest and heaviest bone in the body

A. Divisions

1. *Head—fits into the acetabulum and labrum*
2. *Neck—attached to the capsule of the hip joint—separated from the shaft by the trochanteric line and crest*

   Neck fractures are intra-articular but may be intertrochanteric (more common) or subtrochanteric. Fractures to the neck of the femur may damage the blood supply to the head. Alternative supply via the "artery to the ligament in the head of the femur" may not be adequate.

3. *Shaft (Body)—attachment for muscles*

B. Surfaces

1. *Popliteal*
2. *Patellar*

C. Prominences

1. *Trochanters*
   - i. greater trochanter
     - a. attachment (gluteal tuberosity) for gluteal muscles which abduct the thigh
     - b. attachment for muscles which externally rotate the thigh (piriformis, gamelli, and obturator internus)
   - ii. lesser trochanter—attachment for the psoas major which flexes the thigh at the hip
2. *Intertrochanteric crest (line)*
   - i. between greater and lesser trochanters
   - ii. attachment of the iliofemoral ligament which keeps the femur in the acetabulum
3. *Adductor tubercle—on the medial portion of the distal femur—located above the medial epicondyle*
4. *Linea aspera*
   - i. located along the middle ⅓ of the shaft
   - ii. a continuation of the spiral line above and the supracondylar lines below

5. *Epicondyles*
   i. medial—Posterior to the medial epicondyle is an impression for the medial head of the gastrocnemius.
   ii. lateral

6. *Condyles—medial and lateral—at the distal end of the femur. They articulate with the tibia and are surrounded by ridges.*

   Supracondylar fractures may be located inside or outside the capsule. They may result in a popliteal artery injury. Supracondylar fractures and posterior dislocations of the knee may also damage the femoral artery.

II. MUSCLES (456–462, 471)

A. Anterior (knee extensor and thigh flexor) Group (458, 459)—innervated by the **femoral nerve** (except for the psoas)

1. *Pectineus—may be innervated by the obturator nerve—sometimes classified as a medial thigh muscle*
   i. from the pubic ramus to the femur below the lesser trochanter (linea aspera)
   ii. flexes and adducts the thigh—When the lower limb is flexed, it flexes the vertebral column.

2. *Psoas major—innervated by L2–L3*
   i. from the lumbar vertebrae to the lesser trochanter
   ii. flexes the thigh—When the lower limb is fixed, it flexes the vertebral column.

3. *Iliacus*
   i. from the ilium to the lesser trochanter and hip capsule
   ii. flexes the thigh

4. *Sartorius (Tailor's muscle)*
   i. from the anterior superior iliac spine downward and medially to the medial upper surface of the tibia
   ii. flexes the thigh and leg—abducts and externally rotates the thigh

5. *Quadriceps femoris—the rectus femoris and the 3 vastus muscles extend the leg*
   i. rectus femoris from the ilium to the proximal patella
   ii. vastus muscles from the shaft of the femur to the patella

➢ B. Medial (hip adductor) Group (458, 459)—innervated by the **obturator nerve**

1. *Adductor magnus*
    i. from the ischium and pubic arch to the posterior femur
    ii. adducts and extends the thigh
2. *Adductor brevis—from the inferior ramus of the pubis to the linea aspera*
3. *Adductor longus*
    i. from the pubis to the linea aspera
    ii. adducts the thigh and assists in flexion
4. *Gracilis*
    i. from the pubis to the proximal medial tibia
    ii. adducts the thigh and flexes the leg
5. *Obturator externus*
    i. from the obturator foramen to the trochanteric fossa (located at the posterior part of the neck) of the femur
    ii. adducts and externally (laterally) rotates the thigh

➢ C. Abductors of the Thigh (468, 469)—innervated by the **superior gluteal nerve**

1. *Gluteus medius and minimus—from the ilium to the greater trochanter*
2. *Tensor fascia latae (lateral thigh muscle) (abducts, medially rotates, and flexes the thigh)—from the anterior superior iliac spine to the iliotibial tract which is:*
    i. a part of the fascia lata and an insertion for the tensor fascia lata muscle
    ii. a continuation of the gluteal fascia and an insertion for the gluteus maximus

➢ D. External Rotators of the Thigh (469)—innervation via the **sacral plexus**

1. *Piriformis—from the sacrum and sacrotuberous ligament via the greater sciatic foramen to the greater trochanter*
2. *Obturator internus—from the obturator membrane to the greater trochanter*
3. *Gemelli—from the ischial tuberosity to the greater trochanter*
4. *Quadratus femoris—from the ischium to the quadrate tubercle of the femur*

E. Posterior Thigh Muscles (461)—innervated by the **sciatic nerve** (L4-S3)—extensors of the thigh and flexors of the leg

1. *Hamstrings—semimembranosus and semitendinosus*
   i. from the ischial tuberosity to the medial tibia
   ii. extend the thigh and flex the leg
2. *Hamstring—long head of the biceps femoris*
   i. from the ischial tuberosity to the head of the fibula
   ii. extends the thigh
3. *Biceps femoris*
   i. from the ischium and femur to the fibula and tibia
   ii. externally rotates and flexes the leg and extends the thigh

III. FEMORAL TRIANGLE

A. Boundaries—sartorius, adductor longus, and inguinal ligament

B. Contents—femoral vessels and deep inguinal lymph nodes

IV. FEMORAL CANAL (244, 510)

A. Boundaries—cribriform fascia, fascia lata, and femoral veins

B. Openings

1. *Inlet—the femoral ring*

   Femoral hernias go thru the femoral ring to appear in the femoral triangle. They may present as a palpable mass medial to the femoral pulse. If the small bowel is involved, it may become incarcerated.

   i. boundaries—femoral vein and the inguinal, lacunar, and pectineal ligaments
   ii. contents—lymph nodes including Cloquet's node
2. *Outlet—fossa ovalis*

V. ADDUCTOR CANAL (466, 467, 471)—a groove extending from the apex of the femoral triangle to an opening in the adductor magnus

A. Boundaries—adductor longus, adductor magnus, and vastus medialis

B. Contents—femoral vessels

VI. ARTERIES (466, 467, 470, 477)

A. Femoral—begins deep to the inguinal ligament and passes thru the adductor canal (subsartorial) to enter the popliteal fossa. It has branches to the lower abdomen (superficial circumflex iliac, superficial epigastric, and superficial external pudendal).

1. *Profunda femoris (a major supply to the thigh)*
    i. medial and lateral circumflex
        a. retinacular—surrounds the upper end of the femur
    ii. 4 perforating arteries—pass thru the adductor magnus—allow for collateral circulation to the lower limb when the femoral artery is compromised

B. Obturator Artery—a branch of the internal iliac
1. *Adductor branch*
2. *Hamstring branch*
3. *Artery to the ligament in the head of the femur*

VII. FEMORAL VEIN (466)—located medial to the femoral artery—drains muscles and bones of the thigh (deep tissues). Its superficial tributaries drain the skin.

VIII. INGUINAL LYMPH NODES (377–379, 510)—to external iliacs

A. Superficial
1. *Horizontal nodes—lie parallel to the inguinal ligament and superficial to the femoral vessels—drain the external genitalia (except the testes), anus, and lower anterolateral abdominal wall*
2. *Vertical nodes—drain the superficial areas of the thigh. Their afferent lymphatics follow the course of the saphenous veins.*

B. Deep nodes (4–6)—located along the femoral vein. (Cloquet's node lies in the femoral canal.)

IX. CUTANEOUS NERVES OF THE THIGH (502–504, 508, 509)

A. Ilioinguinal (L1)—supplies the anterior scrotum and medial thigh

B. Lateral Femoral Cutaneous (L2–L3)—a branch of the lumbar plexus—enters the thigh at the lateral end of the inguinal canal—supplies the anterolateral thigh

In laparoscopic surgery, injury to the inguinal ligament may produce anterolateral thigh paresthesias.

C. Femoral (L1–L2)—a branch of the genitofemoral—supplies a small area below the midportion of the inguinal ligament (over the femoral triangle)

D. Intermediate and Medial Cutaneous Nerves of the thigh

E. Obturator (L2–L4)—supplies the upper medial portion of the thigh

F. Posterior Femoral Cutaneous—a branch of the sacral plexus—supplies the posterior thigh and popliteal fossa

Knee

I. BONES (475, 476)

A. Condyles of the femur

B. Patella—located within the quadriceps femoris tendon

II. JOINT (473, 474)—The femur articulates with the tibia and patella.

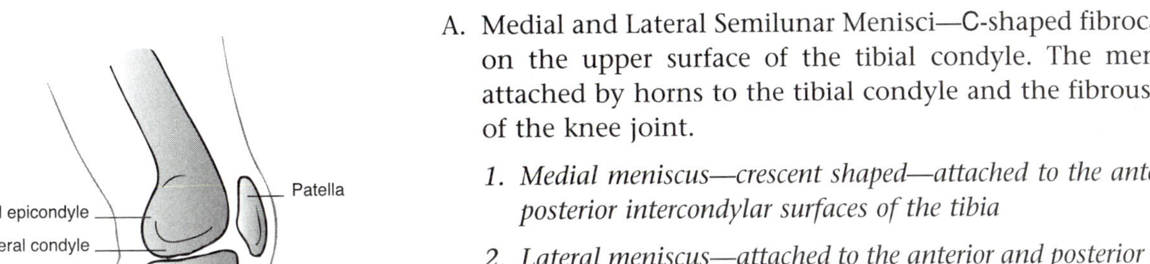

A. Medial and Lateral Semilunar Menisci—C-shaped fibrocartilages on the upper surface of the tibial condyle. The menisci are attached by horns to the tibial condyle and the fibrous capsule of the knee joint.

1. *Medial meniscus—crescent shaped—attached to the anterior and posterior intercondylar surfaces of the tibia*

2. *Lateral meniscus—attached to the anterior and posterior intercondylar areas of the tibia*

B. Articular Surfaces

1. *Condyles and patellar surface of the femur*

2. *Superior articular surface of the tibia*

3. *Articular surface of the patella*

C. Articulations

1. *Medial femoral condyle with the medial tibial condyle and the medial meniscus*

2. *Lateral femoral condyle with the lateral tibial condyle and the lateral meniscus*

3. *Femur with the posterior surface of the patella*

D. Fibrous Capsule

1. *Attachments*

   i. superior—to the femoral condyles and intercondylar fossa

   ii. inferior—to the tibial condyles

E. Anterior Covering—the tendon of quadriceps femoris and the portion called the patellar ligament which extends from the patella to the tibial tuberosity

III. BURSAE (472–474, 476)

A. Main Bursa of the joint

1. *Goes around the cruciate ligaments*

2. *Extends anteriorly and superiorly as the suprapatellar bursa. (It may be entered above the medial border of the patella.)*

B. Bursae Communicating with the joint (deeper bursae)

1. *Popliteus—posterolateral*

2. *Semimembranosus—posteromedial*

3. *Gastrocnemius—adjacent to the medial femoral condyle*

C. Bursae Not Communicating with the joint

1. *Prepatellar—between the skin and lower patella*

2. *Superficial and deep infrapatellar—between the patellar ligament and the tibia*

IV. LIGAMENTS (473–476)

A. Collateral Ligaments (tense during knee extension)

1. *Medial (Tibial)*

i. from the medial epicondyle of the femur to the medial surface of the shaft and the condyle of the tibia

ii. has fibers attached to the medial meniscus

2. *Lateral (Fibular)*

i. from the lateral epicondyle of the femur to the middle of the lateral surface of the head of the fibula

ii. splits the tendon of the biceps femoris muscle

B. Cruciate Ligaments—located within the capsule. They extend from the intercondylar surface of the femur to the anterior or posterior intercondylar notch (eminence). The cruciates cross within the knee joint and connect the femur with the tibia.

1. *Anterior cruciate—extends from the tibial intercondylar surface posteriorly to the lateral femoral condyle (intercondylar notch). It is tense during extension.*

2. *Posterior cruciate—extends from the posterior intercondylar surface anteriorly to the medial femoral condyle (intercondylar notch). It is tense during flexion and prevents posterior movement of the tibia on the femur.*

C. Oblique Popliteal—an extension of the semimembranosus muscle. It strengthens the posterior knee joint.

D. Arcuate Popliteal—from the fibula to the femur and tibia (intercondylar area). It strengthens the posterior knee joint. (The oblique and popliteal ligaments help to prevent hyperextension of the knee.)

E. Patellar—extends from the patella to the tibia and is part of the quadriceps femoris tendon

V. RELATIONSHIPS OF THE KNEE JOINT (472)

A. Anterior—skin, fascia, and quadriceps tendon

B. Posterolateral—biceps tendon

C. Posteromedial—gracilis, semitendinosus tendons, and the sartorius muscle (may be palpated at the superolateral border of the popliteal fossa with the knee flexed)

D. Posterior—popliteal vessels, nerves, and gastrocnemius, plantaris, and semimembranosus muscles

VI. POPLITEAL FOSSA (468)—roughly diamond shaped

A. Borders

1. *Superomedial—semimembranosus and semitendinosus*
2. *Superolateral—biceps femoris*
3. *Inferomedial—medial head of gastrocnemius*
4. *Inferolateral—lateral head of gastrocnemius*

B. Contents of the Popliteal Fossa

1. *The popliteal artery enters the popliteal fossa superomedially and then runs to the middle of the fossa and then inferiorly between the two heads of the gastrocnemius.*
2. *Popliteal vein*
3. *Popliteal lymph nodes are located superficial to the popliteal vessels.*
4. *Nerves in the popliteal fossa*

   i. The common peroneal nerve runs from the superomedial aspect of the fossa to the neck of the fibula.

   ii. tibial nerve

   iii. posterior femoral cutaneous—from the lateral to the medial side and over the popliteal vein

VII. ARTERIES (466, 467, 477)

A. Genicular—a branch of the popliteal artery

B. Descending Genicular—a branch of the femoral artery

C. Anastomosis of A. and B. around the patella

VIII. VEINS (468, 510)

A. Superficial

1. *Great (Greater) (Large) saphenous—runs just posterior to the medial femoral condyle*

2. *Small (Lesser) (Smaller) saphenous*

B. Deep—popliteal vein

IX. LYMPHATICS—run alongside the saphenous vein to enter the popliteal and inguinal nodes

X. NERVES—sensory branches of the femoral, obturator, and common peroneal (common fibular) nerves

XI. KNEE INJURIES

A. **Patella Injuries—Fractures of the patella are usually transverse. Dislocations are usually lateral.**

B. **Ligament Damage—may be associated with cartilage damage. A severe lateral blow may result in medial collateral and medial meniscus tears plus anterior and posterior cruciate ligament tears.**

1. ***Anterior cruciate tear—may be accompanied by effusion***

2. ***Posterior cruciate tear—results from knee hyperextension***

3. ***Collateral ligaments—extra-articular***

C. Arterial Injuries

1. ***The popliteal artery may be injured in a supracondylar fracture of the femur or, more commonly, a posterior dislocation of the knee.***

2. ***The medial head of the gastrocnemius may cause popliteal artery entrapment resulting in an aneurysm and emboli to smaller arteries in the foot (decreased pedal pulses may suggest this diagnosis).***

Leg

I. BONES (478–480)—Tibiofibular ligaments connect the tibia and fibula.

A. Tibia—on the medial side of the leg

1. *Proximal end*
   - i. medial and lateral condyles separated by the intercondylar eminence
   - ii. anterior and posterior intercondylar areas—depressions in front of and behind the eminence for the attachment of the cruciate ligaments
   - iii. The tuberosity of the tibia is for the attachment of the patellar ligament.
2. *Shaft—has 3 borders and surfaces. The upper portion of the shaft is covered by the anterior and lateral muscles.*
3. *Distal end*
   - i. medial malleolus—articulates with the talus
   - ii. fibular notch—lateral to the fibular articulation

B. Fibula

1. *Proximal end—has a synovial tibiofibular joint*
2. *Shaft—connected to the tibia by the interosseous ligament*
3. *Distal end*
   - i. expands to form the lateral malleolus
   - ii. The peroneal (fibularis) longus and brevis muscles pass in the fibular groove located on the posterior border.
   - iii. The medial aspect articulates with the talus.

II. FRACTURES

A. **Tibial plateau—may include the condyles or be displaced**

B. **Tibial shaft—the most frequently fractured long bone in the body**

III. MUSCLES (481–487)

A. Anterior Crural Compartment (484–487)—muscles dorsiflex and invert the foot and extend the toes—all innervated by the **deep peroneal** (anterior tibial) nerve

Overuse may cause periosteal inflammation (shin splints) or Chronic Exertional Compartment Syndrome (increased pressure in the anterior compartment).

1. *Tibialis anterior—located in the medial part of the anterior compartment*

    i. from the lateral condyle and shaft of the tibia to the medial cuneiform and first metatarsal of the foot

    ii. dorsiflexes and inverts the foot

2. *Extensor digitorum longus*

    i. from the lateral condyle of the tibia and the anterior surface of the fibula to the tendons of the lateral 4 toes (terminal phalanges)

    ii. dorsiflexes and everts the foot and extends the toes

3. *Extensor hallucis longus*

    i. from the anterior surface of the tibia to the big toe

    ii. dorsiflexes the foot and extends the big toe

4. *Peroneus tertius*

    i. from the anterior fibula to the 5th metatarsal

    ii. dorsiflexes and everts the foot

B. Posterior Crural Compartment (481–483, 487)—innervated by the **tibial nerve** (L4-S3)

1. *Superficial muscles of the posterior crural compartment innervated by the S1–S2 branches of the tibial nerve. These muscles are attached to the tendo calcaneus. They plantarflex the foot.*

    i. triceps surae—plantarflexors of the ankle and foot

        a. gastrocnemius—plantarflexes the foot and flexes the knee

            1) lateral head—from the lateral condyle of the femur to the tendo calcaneus (Achilles tendon)

            2) medial head—from the medial condyle and the popliteal surface of the femur to the tendo calcaneus (Achilles tendon)

        b. soleus—plantarflexes the ankle

            1) from the fibula, the posterior intramuscular septum, and the soleal line of the tibia to insert with the gastrocnemius

            2) lies under the gastrocnemius

ii. plantaris

a. from the lateral supracondylar line and the popliteal surface of the femur to the tendo calcaneus (Achilles tendon)

b. It is a small muscle which helps the gastrocnemius.

2. *Deep muscles of the posterior crural compartment—innervated by branches of the* ***tibial nerve****—plantarflex and invert the foot*

i. popliteus

a. from the lateral condyle of the femur and the lateral meniscus to the posterior tibia

b. rotates the femur laterally when the tibia is fixed or rotates the tibia medially when the femur is fixed. It flexes and unlocks the knee.

ii. flexor digitorum longus

a. from the posterior surface of the tibia below the soleal line passing behind the medial malleolus and beneath the flexor retinaculum to the distal phalanges of the lateral toes

b. flexes the distal phalanges

iii. flexor hallucis longus

a. from the posterior surface of the fibula to the distal phalanx of the big toe

b. flexes the distal phalanx of the big toe

iv. tibialis posterior

a. from the tibia, fibula, and interosseous membrane to the tuberosities of the navicular and tarsals (except the talus)

b. inverts the ankle and plantarflexes and supinates the foot

➢ C. Lateral Crural Compartment (486, 487)—innervated by the **superficial peroneal (musculocutaneous) nerve** (L4-S1)—dorsiflexes and everts the foot

1. *Peroneus longus*

i. from the lateral condyle of the tibia and the head and lateral surface of the fibula to the lateral aspect of the medial cuneiform and first metatarsal

ii. plantarflexes and everts the foot

2. *Peroneus brevis*

   i. from the lateral surface of the fibula to the tuberosity of the 5th metatarsal

   ii. everts and plantarflexes the foot

IV. MUSCLES WHICH MOVE THE LEG—include the hamstrings, quads, sartorius, gracilis, and gastrocnemius

➢ **V. ARTERIES (477, 483, 485)—The anterior tibial artery supplies the anterior leg, and the fibular (peroneal) artery supplies the lateral leg.**

➢ A. Anterior Tibial (485)—begins at the lower edge of the popliteus muscle and ends in the foot as the dorsalis pedis

1. *Tibial recurrent arteries join* ***anastomoses*** *around the knee.*

2. *Branches to muscles of the anterior crural compartment*

3. *Malleolar branches supply lateral malleolus and* ***anastomoses*** *around the ankle.*

➢ B. Posterior Tibial (483)—passes between the medial malleolus and the posterior tubercle of the calcaneus and terminates beneath the flexor retinaculum by dividing into medial and lateral plantar arteries

1. *Fibular (Peroneal) artery*

   i. supplies muscles of the posterior crural compartment

   ii. supplies muscles of the lateral crural compartment

   iii. is the nutrient artery to the fibula

   iv. **anastomoses** with the arcuate artery on the dorsum of the foot

2. *Circumflex fibular artery—contributes to knee* ***anastomoses***

3. *Nutrient artery to the tibia—****may be injured in a fracture of the tibia and slow down its healing***

➢ **VI. VEINS (508–510)**

A. Superficial Veins

1. *Great (Greater) (Long) (Internal) saphenous vein*

   i. course—begins on the medial side of the foot, passes anterior to the malleolus, and then ascends along the medial aspect of the tibia. The great saphenous goes thru the opening in the fascia lata and enters the femoral vein.

   ii. communications—with the small saphenous and the deep veins which accompany the tibial and peroneal arteries

2. *Small (Lesser) (Short) saphenous vein*

   i. course—begins at the lateral aspect of the dorsal venous arch and ascends along and behind the lateral malleolus and the middle of the back of the leg to enter the popliteal fossa. In the thigh, it enters the femoral vein.

B. Deep veins—accompany arteries

VII. LYMPHATICS (510)

A. Popliteal Node—drains the deep tissues of the leg and foot and the superficial tissues of the lateral leg and foot

B. The Anterior Tibial Node is located in the anterior compartment.

VIII. NERVES (485, 502, 504–509)—The nerves to the leg are derived from the sciatic which divides into tibial and common peroneal nerves in the popliteal fossa. The common peroneal nerve may be palpated approximately 1 inch below the head of the fibula. The common peroneal then branches into the superficial peroneal (musculocutaneous) and deep peroneal (anterior tibial) nerves.

A. Sensory—Cutaneous (502, 504, 508, 509)

1. *Saphenous nerve (a branch of the femoral)—supplies the anterior and medial aspects of the leg*

2. *Sural cutaneous nerves—supply the skin, posteriorly and laterally, of the lower ⅓ of the leg*

3. *Superficial peroneal nerve—supplies the anterolateral portion of the leg*

B. Motor

1. *Tibial nerve (L4-S3)* (505)*—innervates all superficial and deep posterior leg muscles.*

   Injury to the tibial nerve may cause inability to flex the knee and loss of the Achilles reflex.

2. *Common peroneal (common fibular) (lateral popliteal) nerve (L4-S2)* (506) *(curves around the lateral side of the neck of the fibula as it enters the leg)*

   i. superficial peroneal (superficial fibular) (musculocutaneous) nerve—innervates lateral leg muscles and then continues on behind the lateral malleolus

   ii. deep peroneal (deep fibular) (anterior tibial) nerve—innervates anterior leg muscles and the extensor digitorum brevis

   Severe compression of the deep peroneal nerve may cause "foot drop."

Ankle

I. **JOINT—a synovial mortise (hinge joint) formed by the tibiofibular socket and the trochlea of the talus. The ankle joint allows plantarflexion and dorsiflexion of the foot.**

II. **BONES** (478, 488, 489)

A. Tibia—Its distal end forms the medial malleolus.

B. Fibula—forms the lateral malleolus

C. Talus—Its superior articular surface is called the trochlea.

III. **TENDONS** (493)

A. Anterior—extensor muscles (extensor hallucis longus, extensor digitorum longus, anterior tibial, and peroneus tertius)

B. Posterior to the lateral malleolus—peroneus longus and brevis

C. Posterior to the medial malleolus—flexor digitorum longus, flexor hallucis longus, and posterior tibial

D. Posterior—tendo calcaneus (Achilles)

IV. **LIGAMENTS** (491, 492) **(A. and B. stabilize the ankle joint.)**

A. Medial (Deltoid)—a strong ligament which holds the calcaneus against the talus

It is injured in an eversion ankle sprain.

1. *Tibionavicular ligament—from the medial malleolus to the navicular bone*
2. *Talotibial (Tibiotalar) ligament—from the medial malleolus to the talus*
3. *Calcaneotibial ligament—from the medial malleolus to the calcaneus*

B. Lateral (Lateral Collateral)

1. *Anterior talofibular—from the lateral malleolus to the talus*
2. *Posterior talofibular—horizontally from the lateral malleolus to the talus*
3. *Calcaneofibular—from the lateral malleolus to the lateral calcaneus*

C. Anterior—from the tibia to the talus

1. *Transverse crural ligament (also known as the superior extensor retinaculum)*
2. *Cruciate ligament (also known as the inferior extensor retinaculum)*

D. Posterior—from the tibia to the talus

➢ V. RETINACULA (493, 494)

A. Anterior Retinacula—cross over extensor tendons and **combine** to form a lateral attachment to the calcaneus.

1. *Superior extensor retinaculum*
2. *Inferior extensor retinaculum—over the extensor digitorum brevis which is over the extensor hallucis brevis muscle*

B. Lateral (External Annular) Retinacula

1. *Extend from the lateral malleolus to the calcaneus*
2. *Both the superior and inferior peroneal retinacula go over the peroneus longus and brevis muscles on the lateral side of the ankle.*

C. Medial (Flexor) Retinaculum

1. *From the tibial malleolus to the calcaneus*
2. *Extends over the tendons of the flexor muscles, the posterior tibial muscles, the posterior tibial artery, and the tibial nerve*

➢ VI. SYNOVIAL SHEATHS (493, 494)—All tendons crossing over the ankle joint are related to synovial sheaths.

A. Extensor Synovial Sheaths—**deep** to the extensor retinacula

1. *Sheaths of the extensor digitorum longus and hallucis tendons extend ½ inch above and below the inferior retinaculum.*
2. *The sheath of the tibialis anterior tendon extends from above the superior extensor retinaculum to below the inferior extensor retinaculum.*

B. Flexor Synovial Sheaths—**deep** to the flexor retinaculum. They extend approximately 1 inch above and below the retinaculum.

➢ VII. MUSCLES WHICH MOVE THE ANKLE JOINT (482, 484, 494)

A. Dorsiflexors—tibialis anterior, extensor digitorum longus, extensor hallucis longus, and peroneus tertius

B. Plantarflexors—gastrocnemius, soleus, and plantaris

VIII. **ARTERIES—articular arteries from tibial and fibular (peroneal) arteries. [The posterior tibial artery (481) lies posterior to the medial malleolus and beneath the flexor retinaculum.]**

IX. **VEINS—Saphenous veins. (The great saphenous vein (510) passes anterior to the medial malleolus.)**

X. **LYMPHATICS**

A. Which accompany the small saphenous vein—drain deep and lateral superficial tissues

B. Which accompany the greater saphenous vein—drain medial superficial tissues

XI. **NERVES (504, 505)**

A. Tibial nerve—passes behind the medial malleolus (posterior to the tibial artery) to extend to the sole of the foot and divide into plantar nerves to supply the plantar muscles of the foot. The tibial nerve supplies the ankle joint.

B. Sural nerve—lies beneath the apex of the lateral malleolus (accompanied by the small saphenous vein)

XII. **ANKLE INJURIES**

A. **Fractures—They are intra-articular.**

1. ***Classification—Fractures may be classified according to foot position and/or force direction.***

i. **supination—lateral rotation or adduction**

ii. **pronation—lateral rotation**

iii. **impaction**

2. ***Bones—frequently involved***

i. **malleolus—usually accompanied by ligament damage**

ii. **talus—usually a fracture of the neck**

iii. **calcaneus—usually results from a fall**

B. **Ligament injuries—Ligament damage usually presents as a sprain or strain of the *lateral ligaments* as the *result of inversion* of the foot**

Foot

I. BONES (488–490)

A. Tarsus (7)

1. *Posterior row*

i. calcaneus (os calcis)—largest—has 6 surfaces

It may undergo an avulsion fracture due to severe contraction of the tendo calcaneus (Achilles tendon).

ii. talus—2nd largest—between the tibia and calcaneus

The neck may fracture due to severe foot extension.

2. *Middle row—navicular—medial side of the tarsus—located between the talus and cuneiform*

3. *Anterior row*

i. cuboid—on the lateral side, between the calcaneus and 4th and 5th metatarsals

ii. cuneiforms (3)

B. Metatarsus (5)—taper from tarsal to phalangeal end.
The head of the 1st metatarsal presents as a bulge on the medial aspect of the foot.

C. Phalanges (Each proximal phalanx articulates with the head of a metatarsal.)

1. *Big toe (2)*

2. *Other toes (3 each)*

D. Other small accessory and sesamoid bones

II. LIGAMENTS (491, 492, 496)

A. Interosseous Ligaments—support the arches

1. *Plantar calcaneonavicular (Spring ligament)—strong—extends from the calcaneus to the plantar navicular—strengthens the important medial longitudinal arch*

2. *Long plantar—from the bottom of the calcaneus to the cuboid and lateral metatarsals—supports the lateral longitudinal arch*

3. *Short (calcaneocuboid) plantar—supports the lateral longitudinal arch*

4. *Plantar aponeurosis—part of the deep fascia—extends from the calcaneus to the toes (MP joints)*

5. *Talocalcaneal—strong—supports medial and lateral arches*

B. Deep Transverse Metatarsal Ligaments—between the heads of the metatarsal bones

C. Capsular—around the MP joints

III. MUSCLES (494–501)

A. Extrinsic

1. *Anterior—dorsiflex the foot*

   i. from the tibia to the medial cuneiform and 1st metatarsal—tibialis anterior

   ii. from the fibula to the distal phalanges—extensor digitorum longus

   iii. from the fibula to the big toe—extensor hallucis longus

2. *Lateral—evert and plantarflex the foot at the subtalar and transverse tarsal joints*

   i. from the fibula to the metatarsals—peroneus longus and brevis

3. *Posterior*

   i. superficial—plantarflex the foot

      a. from the femur to the calcaneus—gastrocnemius and plantaris

      b. from the fibula to the calcaneus—soleus

   ii. deep—flex the digits and plantarflex the foot

      a. from the fibula to the big toe—flexor hallucis longus

      b. from the fibula to the distal phalanges—flexor digitorum longus

      c. from the tibia and fibula to the navicular and tarsals—tibialis posterior

B. Intrinsic

1. *On the dorsal aspect of the foot*

   i. extensor digitorum brevis—from the calcaneus to the 4 medial toes—supplied by the **peroneal** (anterior tibial) **nerve**

2. *On the plantar aspect of the foot—supplied by plantar branches of the* ***tibial nerve****. The intrinsic muscles strengthen the arches.*

   i. innervated by the medial plantar nerve

a. flexor hallucis brevis—from the cuboid and cuneiform to the proximal phalanx of the big toe

b. flexor digitorum brevis—from the calcaneus and plantar aponeurosis to the proximal phalanges of the big toe

c. abductor hallucis—from the calcaneus to the phalanx of the little toe—strengthens the longitudinal arch

ii. innervated by the **lateral plantar nerve**

a. interossei—aid in flexion

b. lumbricals (4)—from the flexor digitorum longus tendon to the terminal phalanges—aid in flexion

c. flexor digiti minimi brevis—from the base of the 5th metatarsal to the proximal phalanx of the little toe

d. quadratus plantae—from the calcaneus and plantar ligament to the tendons of the flexor digitorum longus—aids in flexion

e. abductor digiti minimi—from the calcaneus to the phalanx of the little toe—strengthens the longitudinal arch

f. adductor hallucis—from the metatarsals and MP ligaments to the proximal phalanx of the big toe

➢ IV. JOINTS (488, 489, 492) (synovial)

A. Subtalar—between the talus and the calcaneus—allows inversion and eversion

B. Transverse tarsal—allows inversion and eversion

1. *Talocalcaneonavicular—ball and socket*
2. *Calcaneocuboid—related to the plantar ligaments*

C. Tarsometatarsal Joints—bases of metatarsals with distal tarsals—allow sliding and gliding

D. Metatarsophalangeal Joints—distal ends of metatarsals with bases of proximal phalanges—permit multiple movements, especially flexion and extension between the metatarsals and the bases of the proximal phalanges

E. Cuneonavicular

F. Cubometatarsal

G. Intermetatarsal

H. Interphalangeal

V. ARCHES

A. Longitudinal—located between the ball of the foot (1st metatarsal) and the calcaneus. The medial and lateral longitudinal arches are supported by extrinsic muscles (especially the peroneus longus).

B. Transverse—formed by metatarsals, cuneiforms, and the cuboid

VI. MOVEMENTS OF THE FOOT

A. Inversion—turns the sole medially

B. Eversion—turns the sole laterally

C. Adduction—turns the toes to the midline of the body

D. Abduction—turns the toes outward

E. Plantarflexion—Moving the front of the foot downward plantarflexes the foot.

F. Dorsiflexion—Moving the front of the foot upward dorsiflexes the foot.

➢ VII. ARTERIES (495, 499, 500)

A. Branches of the Anterior Tibial Artery in the foot

1. *Dorsalis pedis—(Pulsations can be palpated lateral to the tendon of the extensor hallucis longus and 2 inches below the extensor retinaculum.)*

   i. deep plantar

B. Branches of the Posterior Tibial Artery in the foot

1. *Malleolar—to the medial malleolus*
2. *Calcaneus—to the heel*
3. *A branch of the fibular (peroneal) which has* ***anastomoses*** *with the arcuate artery on the dorsum of the foot*
4. *The posterior tibial terminates beneath the flexor retinaculum by dividing into the plantar arteries.*

   i. medial plantar—to the medial side of the foot

   ii. lateral plantar—to the lateral side of the foot—joins the dorsalis pedis to form the plantar arch

➢ VIII. VEINS (508–510)—Superficial Tributaries—drain the skin

A. Great (Greater) Saphenous Vein—begins on the dorsum of the foot and has communications with the lesser (small) saphenous vein

B. Small (Lesser) (Short) Saphenous Vein—begins at the lateral side of the foot (posterior to the lateral malleolus)

➢ IX. NERVES (504–509)

A. Sensory

1. *Common peroneal (fibular) (L4-S2)—branches*

   i. sural (S1)—supplies the skin of the lateral side of the foot

   ii. superficial peroneal (fibular) (L5-S2)—to the skin of the lower anterior leg and foot

   iii. deep peroneal (fibular) (L5)—supplies part of the big toe and the 2nd toe

2. *Femoral branch*

   i. saphenous (S1)—a branch of the femoral—supplies the skin of the medial side of the foot

3. *Tibial branches*

   i. medial plantar (L4-S1)—supplies cutaneous and articular branches to the medial side of the foot

   ii. lateral plantar (S1)—supplies cutaneous and articular branches to the lateral side of the foot

   iii. calcaneal (S1)—supplies the heel

B. Motor Nerves to the Intrinsic Foot Muscles

1. *Branches of the tibial nerve*

   i. medial plantar (S2)—innervates the flexor digitorum brevis and plantar muscles that insert on the big toe

   ii. lateral plantar (S3)—innervates the quadratus plantae, adductor hallucis, abductor digiti minimi, flexor digiti minimi brevis, interosseous muscles, and the lateral 3 lumbricals

2. *Branch of the common peroneal*

   i. deep peroneal (anterior tibial) (deep fibular) (L4)—innervates the extensor digitorum brevis

Questions

Provided below are 16 multiple choice questions for each of the 8 chapters. Annotated answers to all 128 questions begin on page 332.

CHAPTER 1: GENERAL

1. After undergoing surgical resection of a large thyroid nodule, a patient develops tingling around her mouth and muscular spasms. These symptoms are indicative of hypoparathyroidism. One possible explanation for this complication might be surgical damage to the ______________________.

   a. thyroid arteries
   b. parathyroid artery
   c. transverse cervical artery
   d. costocervical trunk

2. A feature that distinguishes the external and internal carotid arteries is that one has branches in the neck while the other does not. Based on your knowledge of the regions that these two arteries supply, you might suspect that if a surgeon has identified a vessel in the neck with branches, she has isolated the ______________ artery.

   a. external carotid
   b. internal carotid
   c. neither

3. All of the following supply blood to the cerebellum and are usually branches of the basilar artery except the ______________ artery.

   a. anterior inferior cerebellar
   b. posterior inferior cerebellar
   c. pontine
   d. All of the above supply blood to the cerebellum and are branches of the basilar artery.

4. A clinical syndrome due to a vascular stenosis exists in which patients classically describe symptoms of vertigo and syncope occurring simultaneously with vigorous exercise of the upper extremities. In this syndrome, blood flows retrograde down the vertebral artery to the axillary artery. Based on your knowledge of vascular anatomy, where might the stenosis that causes this syndrome occur?

   a. at the point where the subclavian artery becomes the axillary artery
   b. in the aorta proximal to the take-off of the subclavian and common carotid arteries
   c. in the vertebral artery just distal to its origin off the subclavian artery
   d. in the subclavian artery distal to the take-off of the vertebral artery
   e. in the subclavian artery proximal to the take-off of the vertebral artery
   f. in the subclavian artery distal to the take-off of the costocervical trunk

5. Studies have indicated that most breast cancers occur in the upper outer quadrant of the breast. Based on the lymphatic drainage of this region, then, you would expect that the lymph nodes most commonly revealing malignancy in these patients are the ______________ nodes.

   a. axillary
   b. supraclavicular
   c. parasternal
   d. internal mammary
   e. anterior mediastinal

6. The cisterna chyli forms at the ____________ vertebral level, marking the ____________ plane.

a. T12; transpyloric
b. L1; transpyloric
c. L1; transumbilical
d. T12; transumbilical
e. L3; transumbilical
f. T12; midsternal

7. Testicular cancer often metastasizes to the ______ nodes because these nodes receive most of the lymphatic drainage from the testes.

a. superficial inguinal
b. deep inguinal
c. internal iliac
d. aortic
e. None of the above receive lymphatic drainage from the testes.

8. In the evaluation of cancer of the anal canal, it is important to know what portion of the canal is involved. This is because lymphatic drainage from the upper portion of the anal canal enters the ______ nodes, while lymphatic drainage from the lower portion of the anal canal drains to the ______ nodes. The anatomical division between the upper and lower anal canal is the ______.

a. internal iliac; superficial inguinal; pectinate line
b. superior mesenteric; inferior mesenteric; pectinate line
c. deep inguinal; superficial inguinal; pectinate line
d. inferior mesenteric; internal iliac; anal verge
e. superior mesenteric; inferior mesenteric; anal verge

9. Ramsay Hunt syndrome is an uncommon though severe complication of infection with the herpes zoster virus. Patients experience ear pain, hearing loss, vertigo, and facial muscle paralysis due to invasion of the virus into the CN VIII ganglion and the geniculate ganglion of CN VII. When the virus is said to invade a ganglion, it has entered a collection of ______.

a. axons
b. dendrites
c. nerve cell nuclei
d. nerve cell bodies in the CNS
e. nerve cell bodies in the peripheral nervous system

10. Which of the following statements correctly explains the difference between a dorsal root and a dorsal ramus?

a. A dorsal root carries only afferent fibers while a dorsal ramus contains both afferent and efferent fibers.
b. A dorsal ramus carries only afferent fibers while a dorsal root contains both afferent and efferent fibers.
c. A dorsal root occurs within the CNS while a dorsal ramus occurs within the PNS.
d. A dorsal ramus occurs within the CNS while a dorsal root occurs within the PNS.
e. The terms are interchangeable.

11. Loss of pelvic splanchnic parasympathetic nerve fibers (S2–S4) may result in ______.

a. urinary incontinence
b. urinary retention
c. absence of peristalsis
d. loss of ejaculatory function
e. a and d
f. b and d
g. a, b, and d
h. all of the above

12. Horner's syndrome is a clinical condition resulting from impaired function of the cervical sympathetic ganglion. Based on your knowledge of the function of this ganglion, which of the following sets of symptoms is consistent with the diagnosis of Horner's syndrome?

a. hyperhidrosis (excessive sweat gland secretion) and pupillary constriction
b. anhidrosis (impaired sweat gland secretion) and pupillary constriction
c. hyperhidrosis and pupillary dilation
d. anhidrosis and pupillary dilation

13. Bilateral damage to the ____________ laryngeal nerves during thyroidectomy may result in permanent vocal cord paralysis. These nerves run in close proximity to the ____________ thyroid artery.

a. external; superior
b. external; inferior
c. superior; superior
d. superior; inferior
e. recurrent; superior
f. recurrent; inferior

14. You are given the remnant of a human skeleton retrieved from an archaeological dig. The specimen is a vertebral process, and you have been asked to identify the region of the vertebral column from which it came. You observe that the transverse processes of this vertebra contain "holes." Based on this information, you conclude that this vertebra is from the ____________________ region.

a. cervical
b. thoracic
c. lumbar
d. sacral
e. coccygeal
f. The region cannot be identified based on this information.

15. You are told that you have a herniated lumbar disk. Based on your knowledge of anatomy, you know that this condition results from herniation of the ____________________ through the ____________________. A surgical procedure that may be proposed to correct this problem is a laminectomy, which involves decompression of the spinal cord by opening the ____________________.

a. annulus fibrosis; nucleus pulposus; anterior vertebral arch
b. nucleus pulposus; annulus fibrosis; posterior vertebral arch
c. nucleus fibrosis; annulus pulposus; anterior vertebral arch
d. nucleus pulposus; annulus fibrosis; anterior vertebral arch
e. nucleus fibrosis; annulus pulposus; posterior vertebral arch

16. A lumbar puncture is usually performed at the L3–L4 or L4–L5 level. This is logical since the purpose of the procedure is to enter the epidural space without damaging the spinal cord. Since the adult spinal cord usually ends at the ____________ level, it is desirable to position the needle below this point. However, one must not place the needle too low because the dural sac typically terminates at the ________ level.

a. L2–L3; S2
b. L1–L2; S2
c. L2–L3; S4
d. L1–L2; S4
e. None of these is accurate.

CHAPTER 2: HEAD

1. A 36-year-old male comes to the emergency room in a coma. His wife states that they were horseback riding. Her husband's horse tripped on a stone and her husband was thrown to the ground. She states that he appeared lucid for a brief interval but then complained of headache, began convulsing, and then became unresponsive. A CT of the head reveals a fractured sphenoid bone and an epidural hematoma. Based on your knowledge of anatomy, you might deduce that the hematoma is the result of injury to the ____________________ artery, a branch of the ____________________ artery.

a. middle meningeal; maxillary
b. occipital; external carotid
c. pterygoid; maxillary
d. superficial temporal; external carotid
e. submental; facial

2. Which of the following is *not* a branch of the internal carotid artery?

a. middle cerebral
b. posterior cerebral
c. posterior communicating
d. anterior cerebral
e. ophthalmic

3. The right and left brachiocephalic veins receive drainage from similar tributaries. One excep-

tion is that on the ____________ side, the brachiocephalic receives the drainage of the thoracic duct.

a. right
b. left

4. Damage to the ____________________ bone might be expected to impair the olfactory nerves.

a. parietal
b. sphenoid
c. ethmoid
d. pterygoid
e. maxillary
f. zygomatic

5. Which, if any, of the following foramina is improperly paired with the structure that travels through it?

a. foramen rotundum; maxillary division of CN V
b. foramen ovale; mandibular division of CN V
c. foramen lacerum; internal carotid artery
d. foramen spinosum; middle meningeal artery
e. jugular foramen; CN IX, X, XI, and internal jugular vein
f. All of the above are correctly matched.

6. A congenital stenosis of the cerebral aqueduct of Sylvius impairs circulation of CSF, causing compression of the brain secondary to hydrocephaly. On CT scan of the head, one would expect to see enlargement of the __________ ventricle.

a. lateral
b. second
c. third
d. fourth
e. callosal

7. Occasionally, head trauma results in herniation (i.e., displacement) of one cerebral hemisphere across the midline. Based on your knowledge of anatomy, underneath what structure must the cerebral hemisphere herniate in order to cross the midline?

a. calcarine fissure
b. falx cerebri
c. foramen magnum
d. tentorium cerebri
e. central sulcus of Rolando

8. The middle meningeal artery lies across the pterion of the skull, so trauma to the pterion may tear this vessel, producing an epidural hematoma. The pterion is formed by the junction of all of the following except the ____________________ bone.

a. frontal
b. parietal
c. temporal
d. occipital
e. sphenoid

9. Damage at the pontomedullary junction of the brain stem is sometimes associated with injury to one of the nerves that innervate the eye muscles. Based on your knowledge of anatomy, the nerve most likely to be injured is CN ________ which innervates the ________ muscle.

a. III; lateral rectus
b. III; superior oblique
c. VI; superior oblique
d. VI; lateral rectus
e. IV; lateral rectus
f. IV; superior oblique

10. The corneal blink reflex requires proper sensory and motor function provided by CN ________________ and ________________, respectively.

a. V; VII
b. V; III
c. III; V
d. III; VII
e. II; III
f. II; VII
g. V (sensory division); V (motor division)
h. None of the above is true.

11. Which of the following is a component, or are components, of the inner ear?
    a. malleus
    b. round window
    c. semicircular canals
    d. utricle
    e. saccule
    f. a and b
    g. a, b, and c
    h. c and d
    i. c, d, and e

12. A 57-year-old male with a long history of heavy pipe smoking develops a malignant squamous cell carcinoma on the central portion of the lower lip. Biopsy of the submandibular nodes reveals the presence of malignancy. Based on your knowledge of the lymphatic drainage of this region, you might assume that the ________ nodes also contain malignancy and that the ________ nodes should be biopsied, since these nodes receive lymphatic drainage from the submandibular glands.
    a. submental; deep cervical
    b. submental; facial
    c. facial; deep cervical
    d. facial; parotid

13. In its mildest form, a congenital cleft of the palate may result in a bifid ________, the posterior end of the soft palate.
    a. palatoglossal arch
    b. palatopharyngeal arch
    c. uvula
    d. epiglottis
    e. levator palatini

14. Sialolithiasis is a condition in which stones form in a salivary duct or gland. In sialolithiasis involving the parotid gland, one might expect to see inflammation at the opening of the parotid duct, located ________.
    a. opposite the first molar
    b. opposite the second molar
    c. opposite the third molar
    d. underneath the tongue

15. The buccinator muscle is a muscle of mastication. It is innervated by CN ________.
    a. V
    b. VII
    c. XII
    d. None of the above

16. A boxer receives a blow to the side of the jaw and falls to the floor. After getting up, he signals to the referee that he cannot close his jaw. Based on your knowledge of anatomy, the boxer is most likely to have ________.
    a. ruptured the external pterygoid muscle
    b. dislocated the temporomandibular joint
    c. fractured the maxilla

CHAPTER 3: NECK

1. A trauma victim comes to the emergency room after a high-speed fall from his motorcycle. He was not wearing a helmet at the time. The cervical spine X-ray series reveals a fracture of the odontoid process. The involved cervical vertebra is ________.
    a. C1
    b. C2
    c. C3
    d. C4
    e. C7

2. The ansa cervicalis is
    a. a loop of cervical nerve roots formed by C1, C2, and C3
    b. a ring of lymphoid tissue formed by the pharyngeal, palatine, and lingual tonsils
    c. a convergence of the inferior cervical ganglion and the first thoracic ganglion
    d. none of the above

3. While isolating the carotid sheath, a surgeon encounters a vascular structure posterior to and outside of the sheath. Based on your knowledge of anatomy, this is most likely to be the ________, a branch of the ________.

a. superior thyroid artery; thyrocervical trunk
b. inferior thyroid artery; thyrocervical trunk
c. superior thyroid artery; external carotid artery
d. inferior thyroid artery; external carotid artery
e. internal thoracic artery; subclavian artery
f. none of the above

4. Which of the following statements regarding the sternocleidomastoid muscle is false?

a. Contraction of the left sternocleidomastoid muscle and relaxation of the right sternocleidomastoid muscle turns the head to the right.
b. It is a landmark separating the posterior and anterior triangles of the neck.
c. The internal carotid artery lies posterior to it in the neck.
d. The external jugular vein lies posterior to it in the neck.
e. All of the above are true.

5. Which of the following muscles of the neck receives partial innervation from a cranial nerve?

a. trapezius
b. sternocleidomastoid
c. platysma
d. digastrics
e. a and b
f. a, b, and c
g. a, b, and d
h. a, c, and d
i. all of the above

6. In order to cannulate the internal jugular vein, one uses the sternal and clavicular heads of the sternocleidomastoid muscle and the clavicle as landmarks. A needle is inserted at the apex formed by the two heads of the muscle. One complication that may arise if the needle is placed too medially in this location is puncture of the ____________________ artery.

a. jugular
b. carotid
c. thyrocervical trunk
d. subclavian
e. inferior thyroid

7. Thyroglossal duct cysts are congenital cysts due to failure of complete migration of thyroid tissue from its origin at the base of the tongue to the normal position of the thyroid gland in the neck. Based on this migration pathway, you should not be surprised that in order to identify and excise the entire tract of the cyst, it is necessary to remove the ____________________ located above the thyroid cartilage and in direct line with the tract of the cyst.

a. cricoid cartilage
b. cricothyroid membrane
c. tracheal cartilage
d. hyocricoid membrane
e. hyoid bone

8. Otitis externa, more commonly known as "swimmer's ear," is a painful infection of the external ear canal. The sensory innervation to this region is supplied by the branches of several nerves. One of these nerves is the ____________________, a branch of C2–C3. This nerve courses through the posterior triangle of the neck and superiorly to the level of the parotid gland, where it separates into anterior and posterior divisions.

a. greater auricular
b. lesser occipital
c. auriculotemporal
d. accessory

9. The structure that forms a loop around the aortic root during its transit through the thoracocervical region is the ____________________.

a. ligamentum arteriosum
b. thoracic duct
c. left phrenic nerve
d. left vagus nerve
e. left recurrent laryngeal nerve

10. Which of the following does not usually drain directly into the brachiocephalic vein?
    a. inferior thyroid vein
    b. superior thyroid vein
    c. thoracic duct
    d. subclavian vein
    e. internal jugular vein
    f. All of the above usually drain directly into the brachiocephalic vein.

11. Which of the following most accurately describes the relations of the thymus?
    a. It is located posterior to the brachiocephalic veins and anterior to the ascending aorta.
    b. It is located anterior to the ascending aorta and posterior to the sternum.
    c. It is located anterior to the prevertebral fascia and posterior to the trachea.
    d. It is located posterior to the trachea and anterior to the esophagus.

12. Swallowing involves a complex process with voluntary and involuntary components. When food or liquid enters the posterior pharynx, a sensory reflex initiates the involuntary steps that lead to closure of the laryngeal aperture so that food is directed away from the airway. Based on your knowledge of anatomy, the sensory arm of this reflex is provided by the ____________ nerve.
    a. facial
    b. mandibular
    c. glossopharyngeal
    d. hypoglossal

13. It is possible to evaluate the presence of impaired swallowing with the aid of videofluoroscopy. The patient is asked to swallow a bolus of contrast-labeled liquid or food and the laryngopharyngeal response is observed on a video screen. One of the expected observations if the swallowing reflex is intact is closure of the laryngeal aperture by the ____________.
    a. vestibular folds
    b. rima glottidis
    c. uvula
    d. epiglottis

14. The orifice of the eustachian tube can become obstructed due to chronic inflammation and enlargement of the lymphoid tissue in the posterior wall of the nasopharynx called the ____________.
    a. adenoids
    b. arytenoids
    c. palatine arch
    d. rima glottidis
    e. palatine tonsils

15. After total thyroidectomy for papillary carcinoma, a patient complains of difficulty breathing and altered phonation. The injury most likely to explain these symptoms is bilateral recurrent laryngeal nerve damage impairing the function of the ____________ muscles.
    a. thyroepiglottic
    b. thyroarytenoid
    c. cricoarytenoid
    d. stylopharyngeus

16. A serious congenital malformation exists in which there is failure of development of the aperture connecting the nasal cavity to the pharynx, resulting in impaired nasal breathing. The structure that normally creates this connection is the ____________.
    a. cloaca
    b. choana
    c. eustachian orifice
    d. rima glottidis
    e. none of the above

CHAPTER 4: THORAX

1. During the repair of a thoracic aortic aneurysm, a surgeon has placed a clamp between the first and second branches of the aortic arch. With the clamp in this location, which of the following arteries is(are) still receiving blood flow?
    a. left common carotid
    b. brachiocephalic
    c. left subclavian
    d. right common carotid

e. a and b
f. b and d
g. a, b, and d

2. Coronary artery bypass grafting is a procedure used to restore blood flow to the heart when the coronary arteries have become stenosed by the presence of atherosclerotic plaques. Sections of artery or vein from other parts of the body are brought to the heart to replace the stenotic vessels. One of the vessels used in this procedure is the left internal thoracic artery, a branch of the ________ artery.

a. left subclavian
b. left common carotid
c. brachiocephalic
d. left internal carotid
e. left external carotid

3. Most of the preganglionic fibers of the parasympathetic system terminate in ________.

a. paravertebral ganglia
b. prevertebral ganglia
c. cell bodies close to the organ they innervate
d. the organ they innervate

4. In an emergency thoracotomy, an incision is made in the intercostal space between the fourth and fifth ribs on the left. Correct placement is critical, so it is important to be able to count the ribs accurately. The sternal angle is a helpful landmark for counting ribs. The sternal angle occurs at approximately the ________ vertebral level, and the costal cartilage of the ________ rib articulates with the sternum at the sternal angle.

a. T2; first
b. T2; second
c. T4; first
d. T4; second

5. Nerves often contain both visceral and somatic afferent fibers, and visceral afferent pain is sometimes referred to the somatic sensory distribution of the corresponding nerve. With this in mind, you might suspect that visceral pain from the diaphragm is referred to the ________.

a. anterior abdominal wall
b. sternum
c. midback
d. shoulder

6. A patient comes to the emergency department after a motor vehicle collision. Chest X-ray reveals a sixth rib fracture but no other injuries. In order to control her pain, you decide to place an intercostal nerve block. You know that the neurovascular bundle lies just ________ each rib, so you place your needle accordingly. Before injecting the anesthetic, however, you aspirate the needle and see bright red blood in the syringe. You conclude that you have entered an intercostal artery. In order to reposition your needle, you should aim slightly more ________.

a. above; superiorly
b. above; inferiorly
c. below; superiorly
d. below; inferiorly

7. Retraction or "dimpling" of the skin overlying the breast can be a visible manifestation of invasive breast carcinoma. It occurs when the breast tissue is pulled toward the pectoral fascia and suggests invasion of the malignancy into which of the following structures?

a. pectoralis major
b. axillary tail
c. lactiferous ducts
d. suspensory (Cooper's) ligaments
e. areola

8. In the setting of a blockage of the superior vena cava above the entrance of the azygous vein, venous drainage from the neck muscles may reach the superior vena cava via the azygous vein and ________ vein, which are connected to each other by way of the ________ vein.

a. hemiazygous; intercostal
b. hemiazygous; accessory hemiazygous

c. accessory hemiazygous; hemiazygous
d. intercostal; hemiazygous
e. intercostal; accessory hemiazygous

9. Cervical mediastinoscopy is a technique that can be used to biopsy the subcarinal lymph nodes in cases of suspected lung carcinoma. During this procedure, a bronchoscope is inserted at the level of the suprasternal (jugular) notch and advanced into the pretracheal fascia to the level of the carina. Based on these landmarks, the bronchoscope extends between approximately what vertebral levels?

a. T2 to T8
b. T2 to T6
c. T2 to T4
d. T4 to T6
e. T4 to T8

10. Select the incorrect statement.

a. The right horizontal and oblique fissures converge at the sixth rib in the midaxillary line.
b. The esophageal hiatus is located at the T10 vertebral level.
c. The left pulmonary artery lies superior to the left bronchus at the root of the lung.
d. The right pulmonary artery lies posterior to the right bronchus at the root of the lung.

11. One of the complications of acute myocardial infarction (AMI) is papillary muscle rupture causing acute ________________. Clinically, this will manifest as a new onset murmur heard loudest at the ____________________________.

a. tricuspid regurgitation; right fifth intercostal space in the midclavicular line
b. mitral regurgitation; left fifth intercostal space in the midclavicular line
c. pulmonic regurgitation; left second intercostal space
d. pulmonic regurgitation; right second intercostal space
e. aortic regurgitation; left second intercostal space
f. aortic regurgitation; right second intercostal space

12. Some patients require long-term central venous access for the deliverance of chemotherapy or parenteral nutrition. To obtain access, a catheter is inserted via the internal jugular or subclavian vein to the junction of the superior vena cava and the right atrium. In this location, the catheter tip is in close proximity to the ________________.

a. SA node
b. AV node
c. fossa ovale
d. coronary sinus

13. During open cardiac surgery, a surgeon wishes to gain control of the superior vena cava. He places his hand in the transverse pericardial sinus. In this location, the superior vena cava will lie ________________.

a. anteriorly
b. posteriorly

14. A 19-year-old male comes to the emergency department with a knife wound to the thorax located at the third intercostal cartilage to the right of the sternum. What structure may the knife have penetrated?

a. aortic arch
b. left brachiocephalic vein
c. superior vena cava
d. right brachiocephalic vein

15. Pleural effusion is an abnormal collection of fluid in the pleural space. It can be caused by several conditions, including pulmonary infections, carcinoma, heart failure, and certain rheumatological diseases. Often, it is necessary to obtain a sample of the fluid for analysis in order to differentiate the possible etiologies. The procedure is called a thoracentesis and is usually performed by inserting a needle in the midaxillary line. In this location, the lung typically ends at the level of the ____________ rib.

a. sixth

b. eighth

c. tenth

d. twelfth

16. Coronary artery bypass graft is the most common surgery in the United States and is indicative of the prevalence of coronary atherosclerotic disease. The coronary artery most commonly involved in atherosclerosis is the anterior descending artery. This artery is a branch of the ______________ artery and travels with the ______________ vein in the anterior interventricular sulcus.

a. right coronary; great cardiac

b. right coronary; anterior cardiac

c. left coronary; great cardiac

d. left coronary; anterior cardiac

CHAPTER 5: ABDOMEN

1. Arterial occlusion at the junction of the aorta and iliac arteries can cause symptoms of thigh and buttock pain, diminished pulses in the lower extremities, and lower extremity muscular atrophy. This junction occurs at approximately the ______________ vertebral level.

a. L1

b. L2

c. L4

d. S2

e. S4

2. A 39-year-old female presents to the emergency department with acute onset abdominal pain and a grossly distended abdomen. She is hypotensive and tachycardic. Diagnostic evaluation reveals erosion of an ulcer in the first portion of the duodenum through the posterior duodenal wall, leading to severe arterial hemorrhage. Based on the anatomy of this region, the artery most likely to be involved is the ______________.

a. right gastro-omental (gastroepiploic)

b. gastroduodenal

c. superior mesenteric

d. right gastric

3. One of the older techniques for managing severe peptic ulcer disease is known as truncal vagotomy. In this procedure, sections of both the right and left vagal trunks are resected at the junction of the esophagus and stomach. This procedure decreases gastric acid secretion by blocking vagal input to the acid-producing cells. One might also suspect that truncal vagotomy would ______________.

a. increase the rate of gastric emptying

b. decrease the rate of gastric emptying

c. increase the heart rate

d. impair swallowing

e. none of the above

4. A surgeon has made an incision in the anterior abdominal wall posterior to the arcuate line. After transecting the rectus muscle, the surgeon will ______________.

a. enter the peritoneal space

b. encounter the aponeurosis of the internal oblique muscle

c. encounter the aponeurosis of the transversus muscle

d. none of the above

5. During development, the testes begin in the abdomen and descend along the course of the gubernaculum. In this pathway, the testes travel through the ______________ inguinal ring, then course ______________ to traverse the ______________ inguinal ring before reaching the scrotum.

a. superficial; medially; deep

b. superficial; laterally; deep

c. deep; medially; superficial

d. deep; laterally; superficial

6. The left renal vein travels in close proximity to the aorta and its major vessels. At one point in its course, the left renal vein is "sandwiched" between two arteries. Occasionally, the vein becomes compressed at this site, causing obstruction to venous drainage. Based on the course of the left renal vein, the two arteries involved in this syndrome are the ______________ ______________ and the ______________.

a. aorta; superior mesenteric artery
b. aorta; inferior mesenteric artery
c. superior mesenteric artery; inferior mesenteric artery
d. aorta; left renal artery

7. A patient presents with a history of abdominal pain and multiple episodes of vomiting which he describes as yellowish-green in appearance. Physical exam reveals high-pitched bowel sounds, and abdominal X-ray demonstrates the presence of air-fluid levels. You suspect obstruction. Blockage at which of the following locations might explain this clinical picture?

a. at the gastric pylorus
b. in the first portion of the duodenum
c. in the small intestine distal to the ligament of Treitz
d. at the common bile duct

8. The Pringle maneuver is a technique used to control bleeding from a traumatic injury to the liver. The surgeon may compress the portal vein and hepatic artery by applying pressure to the ______________.

a. hepatoduodenal ligament
b. falciform ligament
c. ligament of Treitz
d. gastrohepatic ligament
e. gastrolienal ligament

9. Portal hypertension is a potentially life-threatening condition that is most commonly caused by liver cirrhosis. Several shunting procedures have been developed to relieve pressure in the portal venous system in order to manage this condition. Based on your knowledge of venous anatomy, which of the following shunts would *not* be expected to lower pressure in the portal vein?

a. shunt between the splenic and left renal veins
b. shunt between the portal vein and the inferior vena cava
c. shunt between the superior mesenteric vein and the inferior vena cava
d. shunt between the splenic and inferior mesenteric veins

10. The course of the superior mesenteric vessels with relation to the pancreas is complex, but knowledge of the pathway is important when performing surgery in this region. After its origin at the aorta, the superior mesenteric artery lies ______________ to the body of the pancreas and then travels inferior and ______________ to the uncinate process.

a. anterior; posterior
b. posterior; anterior
c. anterior; anterior
d. posterior; posterior

11. When performing a surgical resection of the head of the pancreas, a surgeon must ligate the superior and inferior pancreaticoduodenal arteries that provide the blood supply to the head of the pancreas. Because ligation of these vessels compromises its blood supply, resection of what other structure is required?

a. body of the pancreas
b. gastric pylorus
c. gastric antrum
d. duodenum
e. none of the above

12. Several surface features differentiate the large intestine from the small intestine. For example, the large intestine has three vertical bands of muscle that extend from the cecum to the rectum. These muscular bands are known as ______________.

a. haustra
b. taeniae coli
c. appendices epiploicae
d. rugae

13. Arteries course through the mesentery in order to reach the regions that they supply. Thus, the transverse mesocolon contains the ______________ artery.

a. middle colic
b. right colic

c. ileocolic

d. inferior mesenteric

14. A 17-year-old high school football player has been suffering from generalized malaise and extreme fatigue for several weeks. Nevertheless, insisting that he is feeling better, he is placed in the starting lineup for the final game of the season. While running the ball, he is tackled from the left with a powerful side blow to the bottom of his rib cage. When he arrives in the emergency department, he is hypotensive, tachycardic, and fading in and out of consciousness. What injury is probably responsible for his condition?

a. splenic vein thrombosis

b. splenic rupture

c. hemopericardium

d. diaphragmatic rupture

e. left kidney contusion

15. Pyelonephritis is a painful infection of the kidney. The swelling of the kidney caused by inflammation leads to stretching of the fibrous tissue that covers its surface, producing pain. This fibrous capsule is the ______________.

a. capsule of Bowman

b. capsule of Henle

c. capsule of Gerota

d. medullary capsule

e. renal hilum

16. The lumbar and sacral plexi divide into several nerves that supply the lower abdomen, perineum, and thigh. Which of the following is *not* derived from the sacral plexus?

a. posterior femoral cutaneous nerve

b. lateral femoral cutaneous nerve

c. pudendal nerve

d. nerve to the quadratus femoris

CHAPTER 6: PELVIS

1. Especially in thin people, sitting for extended periods of time can cause pressure on what structure of the bony pelvis?

a. ischial spine

b. pubic tubercle

c. ischial tuberosity

d. coccyx

2. Sciatic nerve irritation can produce pain and paresthesias radiating from the buttock to the lower extremity. This large nerve is subject to injury at various sites along its long course. One of these sites is the greater sciatic foramen. A syndrome has been named to explain chronic buttock pain secondary to sciatic nerve compression by the muscle which crosses this foramen. Based on this information, the name of this condition is most likely ______________ ______________ syndrome.

a. obturator internus

b. obturator externus

c. piriformis

d. iliacus

e. pubococcygeus

3. Anal sphincter continence is maintained by the coordinated actions of several muscles at the anorectal junction. One of the components of maintaining fecal continence involves the ability to alter the angle between the anus and rectum. Contraction of what muscle adjusts this angle to maintain continence?

a. external anal sphincter

b. internal anal sphincter

c. puborectalis

d. iliococcygeus

4. The venous drainage routes of the prostate provide a potential avenue for metastatic spread of prostate cancer. Based on the connections of the prostatic venous plexus, to which of the following locations might the cancer disseminate?

a. penis

b. rectum

c. vertebral column

d. testes

5. Pudendal nerve block is an effective obstetrical anesthetic technique to blunt the pain of episiotomy repair, because the pudendal nerve provides the main sensory innervation to the perineum. Knowledge of the course of the pudendal nerve is necessary in order to place the anesthetic correctly, and the bony pelvis provides several useful landmarks. In its transit between the greater and lesser sciatic foramina, the nerve courses near the ______________________, and hence a needle aimed at this location is likely to hit the nerve.

   a. ischial tuberosity
   b. ischial spine
   c. pubic symphysis
   d. obturator foramen

6. A 24-year-old bicycle racer is injured when the spoke of his wheel becomes caught in the wheel of another cyclist. He straddles over the handles of his bike before falling to the ground. A urethral injury is suspected, and the patient undergoes a retrograde urethrogram. The study demonstrates a tear in the spongy urethra. In what region did the urethrogram demonstrate extravasation of contrast material in order to confirm the diagnosis?

   a. peritoneal cavity
   b. scrotum
   c. prostate
   d. deep perineal space

7. Urethritis is more common in females than in males owing to the proximity of the female urethral meatus to the vaginal and anal orifices. Nevertheless, urethritis does occur in men. Bacteria are especially prone to accumulate in the dilated area at the distal end of the spongy urethra called the ______________.

   a. bulbocavernosus fossa
   b. fossa terminalis (navicular fossa)
   c. prepuce
   d. bulbospongiosus

8. From development to ejaculation, sperm make a complex journey through many structures of the male genital tract. Which of the following accurately sequences the structures through which the sperm travel on this journey?

   a. seminiferous tubules, rete testes, efferent ductules, epididymis, vas deferens
   b. efferent ductules, rete testes, seminiferous tubules, epididymis, vas deferens
   c. rete testes, seminiferous tubules, epididymis, efferent ductules, vas deferens
   d. rete testes, epididymis, seminiferous tubules, efferent ductules, vas deferens
   e. rete testes, efferent ductules, epididymis, seminiferous tubules, vas deferens

9. During an oophorectomy, a surgeon wishes to identify the ovarian vessels. These vessels are located within the ______________________.

   a. ovarian ligament
   b. mesovarium
   c. suspensory ligament
   d. cardinal ligament

10. A physician is performing an open hysterectomy and wishes to ligate the uterine vessels before removing the uterus. She is able to perform the maneuver without injuring the nearby ureter because she is aware of the fact that

   a. the uterine artery passes anterior and superior to the ureter on the lateral pelvic wall
   b. the uterine artery passes posterior and superior to the ureter on the lateral pelvic wall
   c. the uterine artery passes anterior and inferior to the ureter on the lateral pelvic wall
   d. the uterine artery passes posterior and inferior to the ureter on the lateral pelvic wall

11. A 20-year-old female presents to the emergency room with severe lower abdominal pain. She is tachycardic and slightly hypotensive. She reports having noticed a little vaginal bleeding earlier in the day. You are concerned about the possibility of a ruptured ectopic pregnancy with bleeding into the peritoneal cavity. An incision in which location will offer you direct access to the peritoneal cavity in order to sample the peritoneal fluid?

a. posterior fornix of the cervix
b. anterior fornix of the cervix
c. posterior fornix of the vagina
d. anterior fornix of the vagina

12. The pain associated with uterine contractions is transmitted via the

a. pudendal nerve
b. sacral parasympathetics
c. inferior hypogastric plexus
d. uterine plexus

13. A 42-year-old male presents to the emergency department with fever and rectal pain. On exam, you note a fluctuant mass lateral to the anus. You suspect an abscess in the ischiorectal fossa. Which of the following describes the pathway by which this abscess may have formed?

a. Infection in the anal sinuses (crypts) above the pectinate line entered the intersphincteric space; infection spread from the intersphincteric space across the external sphincter to the ischiorectal fossa.
b. Infection in the anal sinuses (crypts) below the pectinate line entered the intersphincteric space; infection spread from the intersphincteric space across the external sphincter to the ischiorectal fossa.
c. Infection in the anal sinuses (crypts) above the pectinate line entered the intersphincteric space; infection spread from the intersphincteric space above the levator ani to the ischiorectal fossa.
d. Infection in the anal sinuses (crypts) below the pectinate line entered the intersphincteric space; infection spread from the intersphincteric space above the levator ani to the ischiorectal fossa.

14. The urogenital diaphragm is a region surrounding the urethra, vagina, and rectum containing two fascial layers, superficial and deep, and two perineal pouches, superficial and deep. Which of the following statements about the superficial perineal pouch is(are) true?

a. It is located below the urogenital diaphragm.
b. It is located between the superficial and deep fascial layers.
c. It contains the deep transverse perineal muscle and sphincter urethrae.
d. It contains the superficial transverse perineal muscle and the central tendon.
e. a and c
f. a and d
g. b and c
h. b and d

15. Micturition requires a complex set of neural events and muscular actions. Which of the following most accurately describes the steps necessary for urination?

a. contraction of the detrusor muscle; relaxation of the external and internal urethral sphincters
b. relaxation of the detrusor muscle; contraction of the external and internal urethral sphincters
c. contraction of the detrusor muscle and external and internal urethral sphincters
d. relaxation of the detrusor muscle and external and internal urethral sphincters

16. A 24-year-old female has just delivered a 9-pound, 5-ounce baby girl. During the delivery, the obstetrician feels that an episiotomy is necessary to enlarge the vaginal canal in order to assist delivery. He chooses to perform the episiotomy by a mediolateral incision extending from the posterior vagina toward the ischiorectal fossa in a line approximately 45° to the midline. The mediolateral incision aims toward what muscle that is important for the maintenance of fecal continence?

a. fourchette
b. levator ani
c. external anal sphincter
d. obturator internus

CHAPTER 7: UPPER LIMB

1. A physician is preparing to perform a thoracentesis. She palpates the inferior angle of the

scapula, then directs her needle into the intervertebral space just below the bone. She is most likely to have entered the intervertebral space between which two ribs?

a. 4 and 5
b. 5 and 6
c. 6 and 7
d. 7 and 8

2. A 49-year-old former professional baseball pitcher experiences chronic shoulder pain secondary to degenerative change in the rotator cuff. This condition is often the result of impingement of a rotator cuff tendon secondary to repetitive shoulder motion. The tendon most at risk for being impinged is the ______ ______ tendon which courses near the subacromial bursa as it blends with the glenoid capsule and inserts of the greater trochanter of the humerus.

a. teres minor
b. teres major
c. supraspinatous
d. infraspinatous
e. subscapularis

3. Superior dislocation of the humerus is rare due chiefly to the strength of the ______ ligament.

a. acromioclavicular
b. coracoacromial
c. coracohumeral
d. glenohumeral

4. The upper limb contains several arterial anastomoses that can protect its blood supply. If the brachial artery is occluded just proximal to the take-off of the profunda brachii artery, through what anastomotic connection might blood flow through the profunda brachii be reestablished?

a. transverse cervical artery
b. subscapular artery
c. circumflex humeral artery
d. ulnar recurrent artery

5. The brachial plexus is a convergence of nerve roots from C5 to T1 that provides innervation to the muscles of the upper limb. Which of the following sections of the plexus are named for their relationship to the axillary artery?

a. trunks
b. cords
c. divisions
d. roots

Refer to the following scenario to answer Questions 6 and 7.
A 10-year-old boy falls from his skateboard and experiences a fracture to the midshaft of the humerus. He is bleeding and is in significant pain. During your exam, you discover that he is still able to extend his forearm but is unable to extend his wrist or fingers.

6. Based on the location of the injury and his clinical exam, the boy is most likely to have damaged the ______ nerve.

a. musculocutaneous
b. radial
c. ulnar
d. median
e. His injuries are not consistent with damage to any single nerve.

7. The artery most likely to be involved in this boy's injury is the ______ artery.

a. profunda brachii
b. brachial
c. axillary
d. radial
e. none of the above

8. "Tennis elbow" is a term used to describe pain incited by the repetitive motion of the backhand tennis stroke. The pain is caused by traction on the extensor tendons at their points of insertion. Based on this information, a more accurate clinical term for this syndrome is ______.

a. medial epicondylitis
b. lateral epicondylitis

c. olecranonitis

d. tuberculitis

9. The ulnar nerve lies relatively superficial in its course around the elbow, and pressure on the nerve can produce an uncomfortable tingling along the sensory distribution of the nerve. This phenomenon is commonly known as hitting one's "funny bone," and is most likely to result from a blow to the ______________ of the humerus.

a. medial condyle

b. lateral condyle

c. medial epicondyle

d. lateral epicondyle

10. Tapping on the biceps tendon in the antecubital fossa produces a reflex jerk due to contraction of the muscle. This reflex is useful in evaluating the function of the ______________ ______________ trunk of the brachial plexus.

a. superior

b. inferior

c. anterior

d. posterior

11. While gathering roses, a gardener pricks her thumb on a piece of rusty wire. There is limited bleeding, and she thinks nothing of the injury. Two days later, her thumb is red and swollen, and movement is restricted and painful. You suspect a synovitis. If the infection remains untreated, to what location may it extend?

a. forearm

b. second digit

c. hypothenar eminence

d. none of the above

12. With the hand held flat in a horizontal plane, movement of the thumb away from the second finger represents what type of motion?

a. abduction

b. adduction

c. flexion

d. extension

13. The function of the lumbricals extends over two joints. It is a combination of ______________ ______________ at the metacarpophalangeal joint and ______________ at the interphalangeal joint.

a. flexion; extension

b. flexion; flexion

c. extension; extension

d. extension; flexion

14. Development of the opposable thumb represents a major advance in the evolution of the human species, allowing for the crucial "pincer grasp." Hence, denervation of the opponens pollicis muscle can be a devastating injury. The nerve that supplies this muscle is the ______________ ______________ nerve.

a. radial

b. ulnar

c. median

d. musculocutaneous

15. Median nerve injury in the flexor retinaculum produces paresthesias in the hand and weakness in the muscles of the thumb. If the median nerve is injured by a supracondylar fracture of the humerus, what additional deficit would you expect?

a. weakness of the extensors of the second and third digits

b. weakness of the flexors of the second and third digits

c. weakness of the flexors of the fourth and fifth digits

d. weakness of the extensors of the fourth and fifth digits

e. none of the above

16. While running to catch a bus, a man falls on his outstretched hand. Subsequently, he develops pain at the base of the thumb on the dorsum of the hand. On exam, he is tender to palpation in the depression formed between the tendons of the abductor pollicis longus, extensor pollicis brevis, and extensor pollicis longus. Based on the history and physical exam, you suspect fracture of what carpal bone?

a. pisiform

b. hamate

c. styloid

d. scaphoid

CHAPTER 8: LOWER LIMB

1. A 78-year-old female slips while getting out of the bathtub. She is brought to the emergency department, where the physician diagnoses an intracapsular fracture of the femoral neck. During her rehabilitation, she begins to complain of a dull aching in the hip, and work-up reveals avascular necrosis of the femoral head. This is a result of disruption of the medial and lateral circumflex femoral arteries, branches of the ____________ artery that normally provide blood supply to this region.

   a. femoral

   b. internal iliac

   c. external iliac

   d. profunda femoris

2. An obese 15-year-old boy complains of intermittent numbness and tingling on the anterolateral aspect of the thigh. You suspect impingement of the ____________ nerve as it crosses the inguinal ligament caused by pressure from the boy's protuberant abdomen.

   a. femoral

   b. femoral branch of the genitofemoral

   c. lateral femoral cutaneous

   d. ilioinguinal

3. Normally, when one leg is raised, the pelvis is stabilized on the contralateral leg due to contraction of the abductors of the thigh. With damage to the nerve supply to the abductors, the pelvis becomes unstable with walking, since this action involves shifting of weight from one leg to the other. This manifests as a characteristic "waddling" or "Trendelenburg gait." Which of the following correctly lists the nerve and muscle(s) involved in this syndrome?

   a. superior gluteal nerve; gluteus maximus muscle

   b. inferior gluteal nerve; gluteus maximus muscle

   c. superior gluteal nerve; gluteus medius and gluteus minimus muscles

   d. inferior gluteal nerve; gluteus medius and gluteus minimus muscles

4. While running toward the end zone, a high school quarterback is tackled by a swiping blow to the lateral aspect of the knee. On exam, you note excess laxity of abduction at the knee joint, suggestive of a medial collateral ligament tear. You are concerned about coexistent damage to the menisci and other knee ligaments, so you proceed with a thorough joint exam. With the patient lying on his back, knee flexed to 90°, and foot flat on the examining table, you grasp the knee joint and gently move it back and forth. You immediately sense anterior displacement of the tibia on the femur and diagnose a tear of the ____________.

   a. anterior cruciate ligament

   b. posterior cruciate ligament

   c. medial meniscus

   d. lateral meniscus

5. A Pott's fracture is a complex ankle injury involving injury to the medial malleolus and the lateral malleolus or distal fibula. This injury occurs when the strong medial deltoid ligament is stretched. Rather than tear away from the bone, the ligament forces a fracture of the medial malleolus. The talus is subsequently pushed laterally, often inducing a secondary fracture of the lateral malleolus or fibula. Based on this information, such an injury most probably occurs due to excess____________ ____________ of the foot.

   a. eversion

   b. inversion

   c. abduction

   d. adduction

6. Knowledge of the course of the great saphenous vein is important to vascular and cardiothoracic surgeons, who often use this vein as a conduit for bypassing arterial occlusions in the arteries of the lower limbs or the heart.

Dissection and removal of the vein is called "harvesting." In the ankle, where would be the best place to look for the great saphenous vein?

a. anterior to the medial malleolus
b. posterior to the medial malleolus
c. anterior to the lateral malleolus
d. posterior to the lateral malleolus

7. Which of the following is a complete list of the bones that participate in the formation of the ankle joint?

a. tibia, fibula, calcaneus
b. tibia, fibula, talus
c. tibia, fibula, calcaneus, talus
d. tibia, talus
e. fibula, talus

8. The act of crossing one's legs involves simultaneous flexion of the thigh and the leg. A muscle capable of performing both of these motions is the ______________________ muscle.

a. gracilis
b. long head of the biceps femoris
c. sartorius
d. semimembranosus

9. The adductor canal, also called Hunter's canal, is a frequent site of occlusion of the femoral artery. This landmark is formed by what three muscles?

a. adductor magnus, adductor longus, vastus medialis
b. adductor magnus, sartorius, gracilis
c. vastus lateralis, vastus medialis, adductor longus
d. rectus femoris, adductor longus, pectineus

10. Assessment of peripheral pulses is an important component of the vascular exam. During such an evaluation, you palpate a pulse near the medial malleolus. However, you are unable to palpate the dorsalis pedis pulse. Based on this information, you are concerned about an occlusion of the______________ ______________ artery.

a. peroneal
b. posterior tibial
c. anterial tibial
d. popliteal

11. In order to be able to support the weight of the body, the knee joint is capable of "locking" the femur on the tibia. In order to initiate walking, the knee must be "unlocked." The muscle normally responsible for this action is the ______________________ muscle.

a. pectineus
b. peroneus
c. popliteus
d. plantaris

12. Prolonged kneeling may cause inflammation of the bursa between the patellar ligament and the tibia. This bursa is called the______________ ______________ bursa.

a. prepatellar
b. suprapatellar
c. infrapatellar
d. main

13. The interossei muscles in the foot are analogous to those in the hand. Thus, as in the hand, the palmar interossei are responsible for______ ______________ of the toes.

a. abduction
b. adduction
c. flexion
d. extension

14. A 52-year-old male comes to the office complaining of pain in the lower back. During the neurological exam, you note that the patient is unable to stand on his tiptoes. The symptoms and exam suggest compression of which nerve root?

a. L3
b. L4
c. L5
d. S1
e. S4

15. The "hamstrings" are muscles in the posterior compartment of the thigh that are involved in knee flexion and thigh extension. With the knee partially flexed, their tendons are prominent at the knee joint. The tendons visible on the medial aspect of the knee represent those of which muscles?

   a. quadriceps femoris and rectus femoris
   b. vastus medialis and vastus lateralis
   c. semimembranosus and semitendinosus
   d. rectus femoris and vastus medialis

16. Knowledge of the anatomy of the femoral triangle is useful when attempting to cannulate the femoral vein. During this procedure, you palpate a strong femoral arterial pulse. In order to pierce the vein, you position the needle just __________ to the point of strongest pulsation.

   a. medial
   b. lateral
   c. superior
   d. inferior

Answers

CHAPTER 1: GENERAL

1. **a.** The inferior thyroid artery provides blood supply to the inferior thyroid gland and the parathyroid glands. The superior thyroid artery is the first branch of the external carotid artery and supplies the superior thyroid gland and the parathyroid glands. The transverse cervical artery is, like the inferior thyroid, a branch of the thyrocervical trunk; however, it supplies blood to the neck muscles of the posterior triangle. The costocervical trunk is a more distal branch of the subclavian artery. The parathyroid artery is not a named vessel.

2. **a.** The external carotid artery supplies the head and neck; therefore, it is not surprising that it has eight named branches. The internal carotid artery is responsible for supplying blood to the brain, so it does not branch until much further up in the head.

3. **b.** The posterior inferior cerebellar artery is generally a branch of the vertebral artery that comes off before the vertebrals converge to form the basilar artery. Both the anterior inferior cerebellar artery and the pontine artery are branches of the basilar artery that contribute blood to the cerebellum.

4. **e.** Subclavian steal syndrome is due to a stenosis of the subclavian artery proximal to the take-off of the vertebral artery. During exercise, the upper extremities' oxygen requirements increase, so the extremities demand more arterial blood flow. A proximal stenosis in the subclavian artery limits the axillary artery's ability to supply blood to the arm. (Recall that the axillary artery is nothing more than a continuation of the subclavian artery in the arm.) To compensate, blood flows retrograde through the vertebral artery and down the subclavian to the axillary artery. In this way, blood in the vertebral artery, which normally supplies portions of the brain including the cerebellum—the organ chiefly responsible for balance—is "stolen" by the arm so that it may be exercised. If the stenosis were distal to the take-off of the vertebral artery, blood flowing retrograde from the vertebral would not be able to cross the stenosis in order to reach the axillary artery.

5. **a.** The axillary nodes receive the chief lymphatic drainage of the breast and most commonly show malignancy in cases of breast cancer originating in the upper outer quadrant of the breast. A lesion in the medial aspect of the breast is more likely to involve the more centrally located internal mammary (also called the parasternal) or anterior mediastinal nodes, although even in these cancers the axillary nodes are more frequently involved. All of these nodes are considered to be part of the regional lymphatic drainage of the breast. Involvement of the supraclavicular nodes in breast cancer, however, is considered to signify distant metastasis according to the current staging system. This is because the supraclavicular nodes are not part of the primary lymphatic drainage of the breast.

6. **b.** The transpyloric plane is at the L1 level and is defined as a horizontal plane approximately halfway between the jugular notch and the pubic symphysis. The transumbilical plane is at the L3–L4 level and is marked by a horizontal plane across the umbilicus. Unlike the other planes listed, the midsternal plane is a *vertical* plane through the sternum. The cisterna chyli marks the beginning of the thoracic duct and is formed to the right of the aorta at the L1 level.

7. **d.** Regional lymphatic drainage pathways commonly follow an organ's arterial blood supply. As such, it is not surprising that the aortic nodes are often sites of metastatic testicular cancer because the right testicular artery is a direct branch of the aorta and the left testicular artery is a branch of the left renal artery, which in turn originates from the aorta.

8. a. The canal is embryologically derived from two structures: the hindgut superiorly and the proctodeum inferiorly. The division between these two regions is the pectinate line. This line is anatomically marked by the inferior edges of anal valves present on the interior surface of the anal canal. Because of this dual embryological origin, the regions above and below the pectinate line have different blood and nerve supplies as well as different lymphatic drainage pathways. The lymphatic drainage from the anal canal above the pectinate line is to the internal iliac nodes, while that of the area below the pectinate line is to the superficial inguinal nodes.

9. e. A ganglion is a collection of nerve cell bodies in the peripheral nervous system. The name for a similar collection of nerve cell bodies in the CNS is a nucleus. A group of axons defines a nerve. Several dendrites normally enter a nerve cell body in order to provide afferent input.

10. a. The difference between a root and a ramus is an important yet subtle distinction. Roots carry only one type of nerve fiber; hence, dorsal roots carry afferent fibers while ventral roots carry efferent fibers. In contrast, dorsal and ventral rami contain both dorsal and ventral roots; hence, rami contain both afferent and efferent nerve fibers.

11. f. The pelvic splanchnics (S2–S4) carry parasympathetic nerve fibers responsible for contraction of the detrusor muscle of the bladder body and relaxation of the urinary sphincter muscle that allows for urination. In addition, the pelvic splanchnics supply parasympathetic fibers to the ejaculatory ducts and seminal vesicles allowing for the constriction that produces ejaculatory fluid. Loss of pelvic splanchnic activity is therefore likely to cause problems with urinary retention (due to loss of inhibition of the urinary sphincter) and loss of ejaculatory function (due to loss of constriction of the ejaculatory ducts and seminal vesicles).

12. b. Horner's syndrome is classically described by the triad "ptosis, miosis, anhidrosis." The clinical picture results from loss of sympathetic input to the eye and sweat glands, leaving parasympathetic function uninhibited. Thus, patients have impaired secretion from the sweat glands, causing anhidrosis. In addition, since the pupil is receiving only parasympathetic input, it constricts, causing miosis. Fibers from the superior cervical ganglion also contribute to the muscular function of the superior tarsus, a muscle that normally contributes to eyelid elevation. Thus, loss of the cervical ganglion occurs, causing a subtle ptosis (eyelid drooping).

13. f. Several branches of the vagus nerve lie in close proximity to the arteries supplying the thyroid, and as such are at risk during thyroid surgery. The external laryngeal nerve is a branch of the superior laryngeal nerve that lies in close proximity to the superior thyroid artery. The external laryngeal nerve sends fibers to the cricothyroid muscle; damage to the nerve may produce vocal weakness and hoarseness. The recurrent laryngeal nerve arises from the vagus, where it crosses the subclavian artery. The recurrent laryngeal supplies all of the laryngeal muscles except for the cricothyroid. The recurrent laryngeal usually lies close to the inferior thyroid artery. Bilateral damage to the recurrent laryngeal nerves results in vocal cord paralysis. In addition, since the recurrent laryngeal nerve innervates the muscles that abduct the vocal cords to allow for air entry, bilateral damage to these nerves causes serious airway compromise that can result in a surgical emergency.

14. a. The "holes" in this vertebra are the transverse foramina that allow for passage of the vertebral arteries, the branches of the subclavian arteries that contribute to the posterior circulation of the brain. Another distinguishing feature of cervical vertebrae is their bifid spinous processes. (Note that these features are for the "typical" cervical vertebrae. C1 contains no spinous process and no body. The dens of C2 takes the place of the C1 vertebral body. In addition, C7 does not contain a bifid spinous process.)

15. b. The vertebral foramen which houses the spinal cord is formed by the vertebral bodies and pedicles anteriorly, and the laminae poste-

riorly. Due to degenerative changes, the nucleus pulposus, the gelatinous substance inside the vertebral bodies, herniates through the annulus fibrosis, the fibrous substance surrounding the nucleus pulposus. This usually occurs in a posterolateral direction and results in compression of the spinal cord, causing pain. If conservative measures fail to relieve the pain, a laminectomy may be offered. The rationale for this procedure is to relieve the pressure being exerted on the spinal cord by opening the posterior vertebral arch.

16. **b.** The adult spinal cord usually ends at the L1–L2 level, while at birth the spinal cord extends slightly further to the L2–L3 level. The dural sac usually terminates at the S2 level. Thus, in order to reach the dural space most safely, one best places the needle between these two points.

CHAPTER 2: HEAD

1. **a.** This patient presents with a classic history for an epidural hematoma caused by bleeding from the middle meningeal artery, a branch of the maxillary artery, which in turn comes off of the external carotid artery. The typical presentation is of a patient who experiences head trauma followed by a "lucid interval" but then a rapid decline into coma. All of the other answer pairs describe correct anatomical branching patterns but would not produce the symptoms present in this patient.

2. **b.** The blood supply to the brain is commonly divided into anterior and posterior circulations. The anterior circulation arises from the internal carotid and includes its major branches: the anterior cerebral, middle cerebral, and posterior communicating arteries. The ophthalmic is also a branch of the internal carotid that supplies the terminal arterial supply to the eye. The posterior circulation of the brain is provided by the vertebrobasilar system which includes the posterior cerebral branch of the basilar artery.

3. **b.** The thoracic duct drains into the left brachiocephalic vein, while the right lymphatic duct drains into the right brachiocephalic vein.

4. **c.** The ethmoid bone forms most of the nasal roof and contains the cribriform plate, a structure through which olfactory fibers travel between the CNS and the olfactory canal. The parietal bone is located on the top and sides of the cranium. The sphenoid bone separates the cranial cavity from the orbit; it contains the pterygoid plates. The maxillary bone forms the front of the face and upper jaw while the zygomatic bone forms the cheeks and a portion of the outer wall of the orbit. While the zygomatic bone contributes to the nasal bridge and other structures of the nasal cavity, the ethmoid bone provides passage for the nerves of olfaction. Therefore, an injury to the ethmoid is more likely to impair olfaction.

5. **f.** All of the answers are correctly matched.

6. **c.** The cerebral aqueduct of Sylvius allows for the circulation of CSF from the third ventricle to the fourth ventricle, so a congenital stenosis of the cerebral aqueduct leads to dilatation of the third ventricle. The lateral ventricles are connected to the third ventricle via the foramina of Monro, so a stenosis in these foramina might be expected to dilate the lateral ventricles. Similarly, stenosis of the lateral foramina of Luschka and the median foramen of Magendie would be expected to cause dilatation of the fourth ventricle, because these foramina are responsible for circulation of CSF from the fourth ventricle to the arachnoid space and the cisterna magna, respectively. There is no second or callosal ventricle.

7. **b.** A subfalcine herniation is a herniation of one cerebral hemisphere across the midline underneath the falx cerebri, the dural septum located between the cerebral hemispheres. The tentorium cerebelli (not cerebri) is another dural septum, but it is located between the occipital lobes and the cerebellum. The calcarine fissure (or sulcus) is a landmark on the medial aspect of the occipital lobe of the cerebrum. The central sulcus of Rolando is a cerebral landmark separating the motor and sensory areas of the cerebrum. The foramen magnum is the aperture through which the spinal cord traverses. The brain can herniate down the foramen magnum, but this type of

herniation does not involve the cerebral hemispheres crossing the midline.

8. **d.** The pterion is formed by the frontal, parietal, temporal, and sphenoid bones. When looking at the lateral aspect of the skull, the pterion is an H-shaped junction formed by these bones. The occipital bone makes no contribution.

9. **d.** CN VI, the abducent nerve, is the most frequently injured nerve supplying innervation to the eye due to its long peripheral course in the brain. The nerve courses near the pontomedullary junction on its journey to the cavernous sinus and ultimately out the superior orbital fissure. Thus, the nerve may be injured in a CNS lesion to this part of the brain stem. The clinical picture associated with injury to CN VI is "lateral gaze palsy," in which the patient is unable to abduct the eyeball past the midline. The patient will complain of diplopia made worse by looking to the far right or far left. CN VI travels in close proximity to CN VII at the pontomedullary junction; thus, lesions of CN VII commonly occur in association with CN VI damage at this region. The other cranial nerves that supply the eye muscles are CN III and CN IV. CN IV provides innervation to the superior oblique muscle. CN III provides innervation to all other extrinsic muscles of the eyeball (superior, inferior, and medial recti and inferior oblique muscle). Neither of these nerves exits near the pontomedullary angle. However, all of these nerves traverse the cavernous sinus, and thus pathology in this region may damage any of these nerves.

10. **a.** The ophthalmic division of CN V provides sensation to the cornea necessary for the afferent arm of the corneal blink reflex (i.e., discomfort). The efferent arm of the reflex (i.e., blinking) requires motor innervation to the orbicularis oculi which is provided by CN VII.

11. **i.** The malleus and round window are structures of the middle ear, along with the incus, stapes, and oval window. The inner ear is comprised of the semicircular canals, the cochlea, and the utricle and saccule, both of which are components of the vestibule.

12. **a.** The submental nodes drain the central portion of the lower lip and send efferent drainage to the submandibular nodes. Thus, if malignancy has already spread to the submandibular nodes, it is likely that the submental nodes also contain malignancy. In the further work-up of this cancer, it would be important to biopsy the deep cervical nodes, because these nodes receive drainage from the submandibular nodes. The facial nodes drain portions of the eyelids, cheeks, and nasal skin and mucosa. While these nodes send efferent drainage to the submandibular nodes, they are not in the pathway of lymphatic drainage for the lower lip and hence would not be expected to be involved in this cancer. Like the submandibular nodes, the parotid nodes drain to the deep cervical nodes; however, the parotid nodes receive primary lymphatic drainage from the nose, eyelids, forehead, and tympanic cavity, not the lips. Because lymphatic drainage from the lips extends into the deep cervical nodes, surgery to treat malignancy in this region not uncommonly involves a radical neck dissection to expose this lymphatic pathway.

13. **c.** The posterior end of the soft palate is the uvula, and a bifid uvula is one manifestation of a congenitally cleft palate. The palatoglossal and palatopharyngeal arches form the soft palate and may also fail to fuse completely in more complex congenital malformations of the palate. Cleft palate may occur in isolation, accompanied by cleft lip, or as part of a complex of malformations (e.g., Pierre-Robin syndrome, a condition in which neonates are born with cleft palate, mandibular hypoplasia, and other defects of the eye and ear). The levator palatini is a muscle that elevates the soft palate. The epiglottis is a structure of the larynx, unrelated to the palate.

14. **b.** The opening of the parotid duct is opposite the second molar. Because of its anatomical proximity to the tooth, inflammation of the parotid caused by infections (e.g., mumps) or by obstructing stones may cause tooth pain.

15. **b.** Although all other muscles of mastication are innervated by CN V, the buccinator muscle

is unique in that it receives innervation from the facial nerve, CN VII.

16. **b.** A blow to the jaw, particularly while the mouth is partially opened, may cause an anterior dislocation of the temporomandibular joint. Such dislocation is also possible due to an acquired weakness of the temporomandibular ligament which normally contributes to the fibrous capsule that surrounds and stabilizes the joint. A muscle rupture is an unlikely injury and does not correlate with the boxer's complaint. Finally, a blow to the jaw may be associated with fracture of the mandible, either in isolation or in conjunction with temporomandibular joint dislocation; however, the punch sustained in this scenario is not likely to have fractured the maxilla.

CHAPTER 3: NECK

1. **b.** The odontoid process is a unique feature of C2, also known as the axis. The odontoid process articulates with the axis (C1) superiorly, which has no vertebral body.

2. **a.** The ansa cervicalis is a loop formed by cervical nerve roots from C1, C2, and C3. The ring of lymphoid tissue formed by the pharyngeal, palatine, and lingual tonsils is called Waldeyer's ring. When the inferior cervical ganglion unites with the first thoracic ganglion, it is called the stellate ganglion. The ansa cervicalis provides innervation to the strap muscles.

3. **b.** The inferior thyroid artery and the thoracic duct usually lie posterior to the carotid sheath in the neck. The inferior thyroid artery is a branch of the thyrocervical trunk off of the subclavian artery. The superior thyroid artery is the first branch of the external carotid artery and would typically not be found this low in the neck. The internal thoracic artery is a branch of the subclavian artery that travels caudally next to the sternum.

4. **d.** The external jugular vein runs across the sternocleidomastoid between it and the platysma. By contrast, the internal jugular vein lies posterior to the sternocleidomastoid muscle in close proximity to the carotid artery. The rest of the statements are correct.

5. **i.** The anterior belly of the digastric is innervated by the mylohyoid nerve, a branch of the inferior alveolar nerve, which is a motor fiber of CN V. The posterior belly of the digastric is innervated by CN VII. The platysma also receives innervation from a cervical branch of CN VII. The trapezius and sternocleidomastoid muscles are innervated by the spinal root of the accessory nerve, CN XI.

6. **b.** The carotid artery lies just medial to the internal jugular vein in this location, so during cannulation of the internal jugular vein it is best to aim laterally. The other arteries are relatively protected during this approach to the internal jugular vein.

7. **e.** The hyoid bone lies superior to the thyroid cartilage. The two structures are connected by the thyrohyoid membrane. In order to perform a complete excision of the tract of a thyroglossal duct cyst, it is necessary to remove the hyoid bone. The cricoid cartilage and cricothyroid membrane are located below the thyroid cartilage. The tracheal cartilage and hyocricoid membrane are fictitious structures.

8. **a.** The greater auricular nerve arises from fibers of C2–C3 and can be found in the posterior triangle of the neck in close proximity to the sternocleidomastoid muscle and external jugular vein. Several other nerves innervate the skin around the ear and external auditory canal, including the lesser occipital nerve and the auriculotemporal nerve. The lesser occipital is also a branch of the cervical plexus but is not found in the posterior triangle of the neck. The auriculotemporal nerve is a sensory fiber from the mandibular branch of CN V. The accessory nerve (CN XI) is a motor nerve that innervates muscles of the pharynx and larynx, as well as the sternocleidomastoid and trapezius muscles; it has no sensory fibers.

9. **e.** The left recurrent laryngeal nerve is a branch of the vagus nerve that loops underneath the aortic arch behind the ligamentum arteriosum, the remnant of the artery that shunted blood away from the lungs to the systemic circulation during fetal life. The left vagus nerve continues inferiorly to the esopha-

geal plexus, where it joins with the right vagus. The vagus nerve is in close association with the phrenic nerve along this course, passing just posterior to it. The thoracic duct travels up from the cisterna chyli just anterior to the thoracic vertebrae. In the thoracocervical region, it loops over the subclavian artery and courses posterior to the carotid sheath, anterior to the vertebral artery, and lateral to the esophagus before draining into the venous system.

10. **b.** The superior thyroid vein typically drains into the internal jugular vein rather than directly into the brachiocephalic vein. The inferior thyroid vein and the thoracic duct drain directly into the brachiocephalic vein, which is formed by the convergence of the subclavian and internal jugular veins.

11. **b.** The thymus is a collection of lymphoid tissue in the superior mediastinum. It is located posterior to the sternum and sternohyoid muscles and anterior to the ascending aorta, brachiocephalic veins, pericardium, trachea, esophagus, and prevertebral fascia. The thymus is most prominent in childhood and gradually atrophies with age.

12. **c.** The glossopharyngeal nerve (CN IX) is responsible for initiating the involuntary phase of swallowing when the presence of food or liquid in the posterior pharynx is sensed. Branches of the vagus nerve (CN X) are also involved in this process. The hypoglossal nerve (CN XII) innervates the tongue and plays a role in the voluntary phase of swallowing but not the involuntary phase. Neither the facial nerve (VII) nor the mandibular nerve, a branch of CN V, is involved in the swallowing process.

13. **d.** It is critical that during swallowing the airway is protected in order to prevent choking, regurgitation, or aspiration. Normally, the process of protecting the airway is involuntary; however, older people and patients with neurological compromise sometimes become unable to protect the airway during swallowing and are thus at increased risk of developing lung infections from food, liquids, or gastric contents entering the lungs. This condition is known as aspiration pneumonia and causes substantial morbidity and mortality in this population.

14. **a.** The adenoids comprise the lymphoid tissue located in the posterior wall of the nasopharynx. In young children, the adenoids sometimes hypertrophy secondary to repeated infections, leading to obstruction of the nasal cavity. Among the complications of this condition is blockage of the eustachian tube orifice in the nasopharynx. Such blockage can predispose these children to recurrent episodes of otitis media that may ultimately compromise hearing. The procedure of choice under these circumstances is adenoidectomy. The arytenoids are paired cartilages that contribute to the larynx. The rima glottidis is the opening between the vocal folds and vocal processes of the arytenoid cartilages. The palatine arch is comprised of mucosal folds located on either side of the posterior border of the soft palate. The palatine tonsils are also a collection of lymphoid tissue, but they are located in the oropharynx, not the nasopharynx.

15. **c.** The cricoarytenoid, thyroarytenoid, and thyroepiglottic muscles are all intrinsic laryngeal muscles involved in vocal cord function. The cricoarytenoids are the muscles responsible for abducting the vocal cords to open the glottic aperture and allow for inspiration. The cricoarytenoids are innervated by the recurrent laryngeal branch of the vagus nerve. This nerve is at risk during thyroidectomy, although a bilateral injury as in this case is extremely rare. The thyroepiglottic muscle widens the inlet of the larynx, while the thyroarytenoid muscle relaxes the vocal cords. The stylopharyngeus is a pharyngeal muscle innervated by CN XI.

16. **b.** Choanal atresia is a congenital malformation that causes severe neonatal airway compromise. Infants classically experience respiratory distress partially relieved by crying. Ordinarily, neonates are obligate nose breathers. In the face of choanal atresia, however, the infant is only able to capture a breath through the mouth, as occurs during crying. None of the other structures listed is present in the nasopharynx.

CHAPTER 4: THORAX

1. **f.** The aortic arch has three major branches. The first is the brachiocephalic artery, which divides into the right common carotid and right subclavian arteries. The second branch of the aortic arch is the left common carotid artery, and the third branch is the left subclavian artery. The clamp is located distal to the takeoff of the brachiocephalic artery, so flow through the brachiocephalic and right common carotid arteries is preserved. The left common carotid and left subclavian arteries are distal to the clamp, so there is no flow through these arteries.

2. **a.** The internal thoracic (mammary) artery is a branch of the left subclavian artery. It is the vessel of choice in coronary bypass graft procedures because it has shown the best long-term patency rate. Other conduits used in this procedure include the saphenous vein, gastroepiploic vein, and inferior epigastric vein.

3. **c.** In the parasympathetic system, the preganglionic fibers terminate in cell bodies near the organ they innervate. The postganglionic fibers then enter the organ. By contrast, in the sympathetic system, the preganglionic fibers terminate in either paravertebral or prevertebral ganglia. Postganglionic fibers then course either through gray rami communicans or to one of the ganglia of the sympathetic trunk. Thus, the preganglionic pathway of parasympathetic fibers is relatively long compared with the preganglionic pathway of sympathetic fibers, while the postganglionic pathway of parasympathetic fibers is short compared with the pathway of postganglionic sympathetic fibers.

4. **d.** The sternal angle occurs at approximately the T4 vertebral level, and the costal cartilage of the second rib articulates with the sternum at this point. The first rib is often difficult to palpate because it is positioned underneath the clavicle. The suprasternal (jugular) notch corresponds to the T2 vertebral level. Emergency thoracotomy is a dramatic procedure used only in the management of certain severe traumas when the patient is extremely unstable. It may be used to access the pericardium for the evacuation of a hemopericardium, for control of hemorrhage from an injury to the thoracic aorta, or for open cardiac massage in the setting of cardiac arrest.

5. **d.** The phrenic nerve, comprised of roots from C3 to C5, provides innervation to the diaphragm. The somatic sensory distribution of these nerves is to the shoulder girdle. Thus, pain from diaphragmatic irritation, caused by gallbladder disease or a lower lobe pneumonia, may be referred to this region.

6. **d.** The neurovascular bundle runs just below the surface of the rib. Within the bundle, the most superior structure is the intercostal vein, followed by the artery, then the nerve.

7. **d.** The suspensory (Cooper's) ligaments bind the breast tissue to the skin and pectoral fascia. Invasion of malignancy into the suspensory ligaments results in retraction that causes the breast tissue to be pulled inward. This creates the dimpling sometimes apparent on the chest. Breast tissue lies on top of the pectoral fascia; the pectoralis major lies deep to this fascial layer. The lactiferous ducts exit from each breast lobe to the nipple, providing a route for the expression of milk. The axillary tail is the portion of breast tissue that extends into the axilla. The areola is the circular area of pigmented skin surrounding the nipple that contains sebaceous glands and smooth muscle.

8. **a.** When the superior vena cava is obstructed above the entrance of the azygous vein, blood from the brachiocephalic veins is shunted into the azygous vein on the right side and the hemiazygous vein on the left side. The intercostal vein provides a connection between the hemiazygous and azygous veins. With this type of obstruction, then, all of the venous drainage from the upper part of the body enters the right atrium via the azygous vein.

9. **c.** The suprasternal (jugular) notch is at the T2 vertebral level, and the carina, the site of bifurcation of the trachea into left and right main stem bronchi, occurs at the level of T4. This corresponds to the sternal angle (angle of Louis). T8 marks the level of the caval foramen,

the aperture that transmits the inferior vena cava and right phrenic nerve.

10. **d.** The right pulmonary artery lies anterior to the right bronchus at the root of the lung. All of the other statements are correct.

11. **b.** The papillary muscles are structures unique to the atrioventricular valves (i.e., tricuspid and mitral valves). They are pillars of cardiac muscle connected to the valvular cusps by cords of connective tissue called the chordae tendinae. After an AMI, damaged papillary muscle may rupture from the chordae tendinae, leading to acute valvular insufficiency. Since the left ventricle is more often involved in infarction, the papillary muscles of the mitral valve are more likely to be affected. Clinically, this manifests as a new onset murmur of mitral regurgitation. It is important to know the areas of auscultation corresponding to the four heart valves. The mitral valve is best heard at the left fifth intercostal space in the midclavicular line, the tricuspid over the sixth intercostal space to the right of the sternum, the pulmonic valve at the left second intercostal space, and the aortic valve at the right second intercostal space.

12. **a.** The SA node is located in close proximity to the junction of the superior vena cava and the right atrium. In fact, during placement of a central line, the catheter tip often precipitates temporary arrhythmias. The fossa ovale is located on the interatrial wall and is the remnant of the fetal foramen ovale. The coronary sinus empties into the right atrium near the entrance of the inferior vena cava. The AV node is located in the medial wall of the right atrium just above the opening of the coronary sinus.

13. **b.** The transverse pericardial sinus occurs at the convergence of the visceral and parietal pericardia. The transverse sinus lies posterior to the aorta and pulmonary trunk and anterior to the superior vena cava.

14. **c.** The superior vena cava enters the right atrium at the level of the right third intercostal space and so would be most at risk of having been penetrated by the stab wound described. The right brachiocephalic vein merges with the left brachiocephalic vein underneath the sternoclavicular junction. Thus, these vessels are relatively protected from direct penetration. The aortic arch begins behind the right edge of the sternum at the second intercostal space and then arches posteriorly and to the left. Thus, it would not be entered by a penetrating wound to the right third intercostal space.

15. **b.** When performing a thoracentesis, it is important to know the normal boundaries of the lung and pleural space in order to prevent pneumothorax—i.e., puncture of the lung substance. During normal respiration, the lung generally reaches the sixth rib in the midclavicular line, the eighth rib in the midaxillary line, and the tenth rib in the midscapular line. The pleural space usually extends the length of an additional two ribs at each of these locations.

16. **c.** The anterior descending artery, also called the left anterior descending artery, is a branch of the left coronary artery. It travels with the great cardiac vein in the anterior interventricular sulcus. The anterior descending artery provides blood supply to most of the left ventricle, right ventricle, and interventricular septum. The other branch of the left coronary artery is the circumflex. Shortly after its origin, the circumflex travels to the left and wraps around the heart to its posterior surface, where it supplies the left atrium and portions of the left ventricle. The main branches of the right coronary artery are the marginal and posterior interventricular arteries. The marginal courses on the inferior border of the heart to supply the right ventricle and apex of the heart. The posterior interventricular travels in the posterior interventricular sulcus and supplies the right and left ventricles and portions of the interventricular septum. Coronary veins travel with the coronary arteries. The great cardiac vein travels with the anterior descending artery, the small cardiac vein with the marginal artery, and the middle cardiac vein with the posterior interventricular artery. The anterior cardiac veins are small vessels that drain directly into the right atrium.

CHAPTER 5: ABDOMEN

1. **c.** The junction of the aorta diverges into the right and left iliac arteries at the L4 vertebral level, marked superficially by a horizontal plane through the iliac crests. L1 marks the take-off of the superior mesenteric artery from the aorta, and L2 marks the take-off of the renal arteries. The two other main branches of the aorta are the celiac trunk and the inferior mesenteric artery. The celiac trunk originates at the T12 level and the inferior mesenteric artery at the L3 level.

2. **b.** The gastroduodenal artery lies posterior to the first portion of the duodenum, and so it is most likely to be the source of hemorrhage in the scenario described. The first portion of the duodenum is the most common site for duodenal ulcers. The right gastric artery is a branch of the gastroduodenal which ascends along the lesser curvature of the stomach and anastomoses with the left gastric, one of the three vessels of the celiac trunk. The right gastro-omental (gastroepiploic) is also a branch of the gastroduodenal artery. It courses along the greater curvature of the stomach and anastomoses with the left gastro-omental (gastroepiploic), a branch of the splenic artery. The superior mesenteric artery lies anterior to the third portion of the duodenum.

3. **b.** In addition to severing the vagal fibers that innervate acid-producing parietal cells in the stomach, truncal vagotomy sacrifices vagal input to the pylorus. This impairs vagally induced relaxation of the pyloric channel, leading to a decreased rate of gastric emptying. Thus, truncal vagotomy requires the addition of a gastric drainage procedure to allow for adequate emptying of the stomach. Such procedures include pyloroplasty, which involves destruction of the pyloric sphincter, and gastrojejunostomy, in which the dysfunctional pyloric channel is bypassed and gastric contents are routed directly to the jejunum. A truncal vagotomy would not affect vagal fibers involved in heart rate control or swallowing since the procedure transects the vagi distal to the offshoot of these fibers.

4. **a.** The rectus sheath is formed by the aponeuroses of the external oblique, internal oblique, and transversus muscles. The layers of the sheath differ depending on whether one is above or below the arcuate line. Above the arcuate line, aponeuroses of the external and internal obliques contribute to the sheath in front of the rectus and aponeuroses of the internal oblique and transversus form the sheath posterior to the rectus. Below the arcuate line, however, all three aponeuroses travel in front of the rectus muscle. Thus, the muscle is in direct contact posteriorly with the peritoneal cavity.

5. **c.** The testes descend from the abdomen through the deep inguinal ring which is located approximately midway between the anterior superior iliac spine and the pubic symphysis above the inguinal ligament. After traversing the deep inguinal ring, the testes travel medially to enter the superficial inguinal ring located at the medial aspect of the inguinal ligament. Failure of complete testicular descent results in a condition known as cryptorchidism. Cryptorchid testes are at increased risk for the development of testicular cancer.

6. **a.** In order to reach the inferior vena cava, the left renal vein travels approximately perpendicularly behind the superior mesenteric artery and in front of the aorta. Here, it can be wedged in the angle formed between these two arteries, leading to compression and venous congestion. This condition is known as the "nutcracker" phenomenon. The two arteries are analogous to the two arms of a nutcracker, and the left renal vein is compressed between these two vessels as is a nut between the arms of a nutcracker.

7. **c.** The vomitus described suggests the presence of bile. This places the obstruction distal to the entrance of the common bile duct at the ampulla of Vater in the second portion of the duodenum. An obstruction at the gastric pylorus or in the first portion of the duodenum would not be expected to produce bilious vomiting. In addition, an obstruction at the orifice of the common bile duct would presumably block biliary drainage from the gallbladder and liver into the duodenum. An obstruction distal

to the ligament of Treitz, the junction of the duodenum and jejunum, however, could produce the clinical picture described.

8. **a.** The hepatoduodenal ligament contains the portal triad which is comprised of the portal vein, hepatic artery, and common bile duct. Applying pressure to this ligament thus compresses the two major sources of blood supply to the liver, which may be helpful in controlling bleeding from a liver laceration. The hepatoduodenal, cholecystoduodenal, and gastrohepatic ligaments comprise the lesser omentum. The gastrohepatic ligament connects the lesser curvature of the stomach to the liver. The falciform ligament connects the liver to the anterior abdominal wall and contains the ligamentum teres, a remnant of the fetal umbilical vein. The ligament of Treitz marks the junction of the duodenum and jejunum. Finally, the gastrolienal (gastrosplenic) ligament is part of the greater omentum. It connects the stomach to the spleen.

9. **d.** The portal vein is formed by the union of the superior mesenteric vein and the splenic vein. Thus, any procedure that redirects blood from one of these vessels into the inferior vena cava, bypassing transit through the liver, would be expected to reduce portal hypertension. The left renal vein drains directly into the inferior vena cava, so directing the splenic vein to the left renal vein provides an effective shunt. Similarly, connecting the superior mesenteric vein or the portal vein itself to the inferior vena cava allows blood to bypass the liver. The inferior mesenteric vein drains into the splenic vein before the latter merges with the superior mesenteric vein. Thus, this connection in fact contributes to the normal portal venous blood flow.

10. **b.** The pancreas is divided into four major parts: the head, neck, body, and tail. The head also contains the uncinate process, an extension of tissue that courses slightly superior to and to the left of the head. After leaving the aortic trunk, the superior mesenteric artery travels posterior to the body of the pancreas. It then courses anterior to the uncinate process.

11. **d.** The superior mesenteric artery gives off the inferior pancreaticoduodenal artery which, together with the superior pancreaticoduodenal artery, a branch of the gastroduodenal, provides blood supply to the head of the pancreas and the duodenum. Thus, resection of the head of the pancreas requires concomitant resection of the duodenum. Branches of the splenic artery supply blood to the body and tail of the pancreas. The gastric pylorus and gastric antrum receive blood supply from the gastric arteries.

12. **b.** The taeniae coli are three longitudinal muscular bands that extend the length of the large intestine. Haustra are sacculations of the wall of the colon that protrude between these muscular bands. The appendices epiploicae are fat-filled peritoneal sacculations on the surface of the intestine. Rugae are the deep folds of the stomach.

13. **a.** The middle colic artery, a branch of the superior mesenteric, provides blood supply to the proximal transverse colon and reaches this region by way of the transverse mesocolon. Other branches of the superior mesenteric artery include the ileocolic, which supplies the ileum and cecum, and the right colic, which supplies the ascending colon. The left colic artery is a branch of the inferior mesenteric artery that supplies the distal transverse colon and descending colon. Other branches of the inferior mesenteric artery include the sigmoid artery to the sigmoid colon and the superior rectal artery to the proximal rectum.

14. **b.** The patient's injury and symptoms are worrisome for an acute splenic rupture. His several-week history of malaise and fatigue is suggestive of infectious mononucleosis, a viral infection that can cause splenic enlargement. This condition might have predisposed him to splenic rupture caused by the left-sided tackle he received. The ninth to eleventh ribs overlie and protect the spleen. However, a fracture to any of these ribs may tear the splenic substance. Owing to its extreme vascularity, any injury to the spleen may rapidly cause severe hemorrhage and hypovolemic shock.

15. c. There are many named features of the kidney. The capsule of Gerota is the fascial covering of the kidney that is stretched when the kidney becomes inflamed. The kidney is divided into a peripheral cortex and an inner medullary region. Within the substance of the kidney are corpuscles containing the renal glomeruli and loops of Henle. Bowman's capsule surrounds the renal corpuscle. The renal hilum describes the area where the renal vessels and ureters enter and exit the kidney.

16. b. The sacral plexus divides into the posterior femoral cutaneous nerve (S1–S3), the pudendal nerve (S2–S4), and nerves to several muscles in the pelvis, including the quadratus femoris. The lumbar plexus forms the lateral femoral cutaneous (L2–L3) as well as the genitofemoral (L1–L2), iliohypogastric (L1), ilioinguinal (L1), femoral (L2–L4), and obturator (L2–L4) nerves. Nerve roots from both plexi contribute to the large sciatic nerve (L4-S3).

CHAPTER 6: PELVIS

1. c. The ischial tuberosities are located at the posterior aspect of the ischium. In the seated position, the weight of the torso is transmitted to these bony prominences. Hence, especially in people with little gluteal muscle covering these structures, the ischial tuberosities can become sore after remaining in the seated position for an extended period of time. The ischial spines are sites for the attachment of several pelvic muscles and ligaments but are not under direct pressure. The pubic tubercle is part of the anteriorly located pubis bone. It is the medial attachment site for the inguinal ligament. The coccyx is the product of the fused coccygeal vertebrae. Although the tip of the coccyx is palpable at the superior aspect of the anal cleft, it does not bear weight.

2. c. The piriformis muscle crosses the greater sciatic foramen and divides it into upper and lower portions. The sciatic nerve normally transits the greater sciatic foramen underneath the piriformis, although this course is variable, and the nerve may even cross through the muscle in certain cases. If this is the case, or if the piriformis muscle becomes hypertrophied, it can irritate the sciatic nerve, causing pain and paresthesias from the buttock down the lower extremity in the distribution of this nerve. The iliacus lines the ilium and inserts on the lesser trochanter of the femur. The obturator internus courses through the lesser sciatic foramen. The obturator externus extends from the obturator foramen to the femur. The pubococcygeus is part of the levator ani muscle which maintains the integrity of the pelvic floor.

3. c. The puborectalis muscle inserts on the pubic symphysis and forms a sling around the anorectal junction. Thus, contraction of this muscle is responsible for changing the angle between the anus and rectum. The puborectalis, iliococcygeus, and pubococcygeus form the levator ani muscle, the muscle fibers of which merge with the external anal sphincter. The external anal sphincter is composed of striated voluntary muscle, while the internal anal sphincter is made up of smooth involuntary muscle. Both of these sphincters are crucial to maintaining fecal continence but do not act by altering the angle of the anorectal junction.

4. c. The prostatic plexus drains into the vesical and vertebral plexi and the internal iliac veins. This provides a route for cancer of the prostate to reach the vertebral column, skull, and brain.

5. b. The pudendal nerve crosses the ischial spine on its course from the greater to the lesser sciatic foramen, and thus the ischial spine is a useful landmark for placing a pudendal nerve block. With the patient in lithotomy position, the spine of the ischium can be palpated through the vagina, allowing for correct placement of the needle.

6. b. A tear in the spongy urethra causes urine to extravasate through the superficial perineal space into the scrotum as well as beneath Scarpa's and Colles' fasciae in the anterior abdominal wall. A disruption of the prostatic urethra, which is located above the urogenital diaphragm, allows urine to accumulate in the extraperitoneal space. This is also true with a rupture of the posterior wall of the bladder. These injuries most commonly occur in the context of a coexistent pelvic fracture. The spongy urethra, however, is not uncommonly

damaged in straddle injuries of the type described in which the urethra is compressed against the bony pubis. A retrograde urethrogram is performed by inserting a small volume of contrast material into the urethral meatus. X-rays of the pelvis are then taken to assess the patency of the urethral passage.

7. b. The fossa terminalis (navicular fossa) represents a dilatation at the end of the spongy urethra where bacteria may pool to cause infection. The bulbocavernosus and bulbospongiosus are muscles that comprise the penis. The bulbospongiosus also surrounds the spongy (penile) urethra and contributes to urinary continence. The prepuce is a fold of skin overlying the glans penis. This skin is removed in a circumcision. In uncircumcised men, bacteria can also collect underneath this skin.

8. a. Sperm form in the testes and coalesce in the seminiferous tubules. The channels of the seminiferous tubules form the rete testes which in turn become the efferent ductules. The efferent ductules carry sperm to the epididymis, where they are stored. After exiting the epididymis, they enter the vas deferens, a thick muscular tube that travels in the spermatic cord. The vas deferens joins with the duct of the seminal vesicle behind the bladder to form the ejaculatory duct which enters the prostatic urethra at the seminal colliculus.

9. c. The female reproductive organs are surrounded by several peritoneal, connective tissue, and smooth muscle structures that provide support, transmit vessels, and form connections between the various organs. The suspensory (infundibulopelvic) ligament contains the ovarian vessels and nerves. The mesovarium extends from the anterior portion of the ovary to the broad ligament. Both of these are peritoneal supports. The ovarian ligament connects the ovary to the uterus. The cardinal (transverse cervical) ligament is lower in the female reproductive tract. It connects the cervix to the lateral pelvic wall.

10. a. The relationship between the ureter and uterine artery is partially described by the saying "water goes underneath the bridge." This phrase is meant to emphasize that the uterine artery passes anterior to the ureter on the lateral pelvic wall. In this location, the uterine artery is also slightly superior to the ureter. Careless dissection in this region places the ureter at serious risk of inadvertent ligation.

11. c. The vaginal fornices are recesses of the vagina surrounding the cervix. The posterior fornix is in direct relation to the rectouterine space (pouch of Douglas), the most inferior extension of the peritoneal cavity. Hence, an incision in the posterior vaginal fornix allows for direct sampling of the peritoneal fluid. In the case of a ruptured ectopic pregnancy, one would expect blood in the aspirate. The procedure, known as culdocentesis, can also be used to diagnose and drain pelvic abscesses.

12. c. The visceral afferent innervation of the uterus is transmitted via the inferior hypogastric plexus (T10-L1); hence, it provides the pathway for the pain associated with uterine contractions during the first stage of labor. The sacral parasympathetics (S2–S4) provide innervation to the detrusor muscle, the sphincter urethrae, and other structures of the pelvis. In addition, visceral afferent fibers from the cervix follow the course of these nerve roots. The pudendal nerve provides sensation to the perineum. Since the pudendal nerve does not provide innervation to the uterus, a pudendal nerve block is not effective in controlling the pain associated with the first stage of labor.

13. a. The anal sinuses are located between the anal columns and behind the anal valves above the pectinate line. The sinuses drain into the intersphincteric space. From there, infection originating in the anal sinuses can travel in several directions. The abscess described has entered the ischiorectal fossa. This is a space lateral to the external sphincter. Thus, from the intersphincteric space, the infection must have traveled across the external sphincter to reach the ischiorectal fossa. Had the abscess spread above the levator ani muscles, it would have entered the supralevator space.

14. f. The superficial perineal pouch is located below the urogenital diaphragm and contains,

among other structures, the superficial transverse perineal muscle and the central tendon. The deep perineal pouch is between the two fascial layers of the urogenital diaphragm. Its contents include the deep transverse perineal muscle and the sphincter urethrae.

15. **a.** Micturition is a combined voluntary and involuntary process involving coordinated muscle contraction and relaxation. The parasympathetic system is active during urination, while the sympathetic system contributes to the process of urinary retention. The detrusor muscle is located in the body of the bladder. For stored urine to be released, the detrusor muscle must contract. In addition, the external and internal urethral sphincters must be relaxed to allow an exit for the urinary stream. Thus, urination occurs by the coordinated action of muscle contraction and relaxation.

16. **b.** The mediolateral incision begins at the fourchette of the vagina, the posterior convergence of the labia minora and labia majora. The incision aims directly at the levator ani muscle, which forms the medial boundary of the ischiorectal fossa, and hence this muscle is at risk with this type of incision. The levator ani muscle is important not only for maintaining fecal continence but also for supporting the pelvic floor. Unlike a midline procedure, the mediolateral episiotomy aims away from the external anal sphincter which attaches to the central tendon in the midline. The obturator internus forms the lateral wall of the ischiorectal fossa. It is a lateral rotator of the thigh and does not play a role in maintaining fecal continence. In addition, the incision would need to cross the levator ani and ischiorectal fossa before reaching the obturator internus, so this muscle is not at risk during such a procedure.

CHAPTER 7: UPPER LIMB

1. **d.** The superior angle of the scapula is at the T2 vertebral level. The inferior angle is marked by a plane through the T7 level. Thus, the scapula overlies the second to seventh ribs, so positioning the needle just below the level of the inferior angle of the scapula places the needle in the intervertebral space between the seventh and eighth ribs.

2. **c.** The rotator cuff is formed by the tendons of the supraspinatous, infraspinatous, teres minor, and subscapularis. The supraspinatous tendon lies close to the subacromial bursa as it merges into the glenoid capsule and inserts on the greater tubercle of the humerus. Impingement of the supraspinatous tendon between the acromion and humeral head is a frequent cause of rotator cuff injury. Repeated impingement leads to the formation of osteophytes, or abnormal calcium deposits, within the tendon. These osteophytes can cause irritation of the subacromial bursa, producing a condition called subacromial bursitis. Rotator cuff injury is usually a chronic condition managed by anti-inflammatory medications and physical therapy. Surgical repairs are used in specific cases.

3. **b.** The coracoacromial ligament attaches to the coracoid process and acromion. This strong ligament prevents superior dislocation of the head of the humerus. The glenohumeral joint is weakest in its anteroinferior aspect, so dislocations most commonly occur in this direction.

4. **c.** An occlusion of the brachial artery proximal to the profunda brachii branch obstructs the normal route of blood flow through this artery. However, blood flow through the profunda brachii may be reestablished owing to anastomoses with the anterior and posterior circumflex humeral arteries. These are branches of the axillary artery that wrap around the neck of the humerus. A small branch of the profunda brachii connects with these arteries. Through this connection, blood from the axillary artery might be able to reach the profunda brachii in order to restore blood supply to the upper limb. The transverse cervical artery, a branch of the thyrocervical trunk, and the subscapular artery, a branch of the axillary artery, contribute to the anastomotic network around the scapula. The ulnar recurrent artery is a branch of the ulnar artery. This artery occurs distal to the site of occlusion and thus could not restore blood flow to the profunda brachii.

5. **b.** The lateral, medial, and posterior cords are named for their location with respect to the axillary artery. The roots emerge just distal to the intervertebral foramen. The trunks form in the posterior triangle of the neck, and the divisions form underneath the clavicle.

6. **b.** The radial nerve lies in the radial groove of the humerus and is at risk in a midshaft fracture of this bone. The radial nerve supplies the wrist extensors in the posterior compartment of the forearm as well as the extensor muscles of the four digits. Extension of the forearm is a function of the triceps muscle. Although this muscle is innervated by a branch of the radial nerve, the branch occurs more superiorly in the humerus, explaining why extension of the forearm is preserved in this case.

7. **a.** The profunda brachii, a branch of the brachial artery, travels with the radial nerve in the posterior compartment of the arm. The axillary artery becomes the brachial artery at the insertion of the teres major muscle onto the humerus. The radial and ulnar arteries are branches of the brachial artery that form just distal to the elbow joint.

8. **b.** The extensors of the forearm have a common tendinous insertion on the lateral epicondyle of the humerus, and hence the medical term for "tennis elbow" is lateral epicondylitis. Forearm flexors insert on the medial epicondyle.

9. **c.** The ulnar nerve lies near the medial epicondyle of the humerus. As such, a blow to this site is most likely to hit the ulnar nerve at the elbow joint, producing the characteristic sensory perception.

10. **a.** There are three trunks of the brachial plexus. The superior trunk is formed from the roots of C5–C6, the middle trunk from the C7 nerve root, and the inferior trunk from the roots of C8-T1. The C5–C6 roots innervate the biceps brachii, the tendon of which produces the biceps reflex. Thus, the integrity of the biceps tendon reflex is useful in evaluating the superior trunk of the brachial plexus.

11. **a.** There are five synovial bursae in the hand. The bursae surrounding the tendons of the second, third, and fourth digits extend from the metacarpal joint to the distal phalanx. The ulnar bursa surrounds the fifth digit and continues beyond the flexor retinaculum into the forearm. The bursa surrounding the first digit is called the radial bursa. It too extends up the forearm. Owing to these boundaries, an untreated infection of the bursa around the first digit could extend into the forearm.

12. **d.** It is important to be able to report the various joint motions accurately, and the names given to the movements of the thumb can be somewhat confusing. With the hand held flat in a horizontal plane, movement of the thumb away from the second digit is termed extension, and return of the thumb to its original position is called flexion. By contrast, abduction and adduction are motions of the thumb in a plane perpendicular to the palm. When describing motion of the other digits, however, spreading of the digits in a plane horizontal to the palm is called abduction, and return to the resting position is called adduction. Flexion and extension of the digits are motions of the joint at a perpendicular angle to the palm.

13. **a.** The lumbricals produce flexion at the metacarpophalangeal joint and extension at the interphalangeal joint. As such, they are nicknamed the "bye-bye" muscles in that the net effect of their action is to produce the motion associated with waving goodbye.

14. **c.** The median nerve provides innervation to the opponens pollicis. The majority of the intrinsic hand muscles receive innervation from the ulnar nerve. However, except for the adductor pollicis, the muscles of the thenar eminence are innervated by the median nerve. Neither the radial nerve nor the musculocutaneous nerve innervates intrinsic hand muscles.

15. **b.** The median nerve innervates most of the flexors of the wrist and digits, including the palmaris longus, flexor carpi radialis, flexor digitorum superficialis, and the lateral portion of the flexor digitorum profundus. The medial portion of the flexor digitorum profundus,

which flexes the fourth and fifth digits, is innervated by the ulnar nerve, as is the flexor carpi ulnaris. Hence, damage to the median nerve just above the elbow joint produces weakness of the flexors of the second and third digits with partial sparing of the flexors of the fourth and fifth digits. The median nerve and brachial artery are at risk in supracondylar fractures of the humerus. In fact, if damage to the brachial artery remains untreated, a condition known as Volkmann's ischemic contracture may result. In this condition, ischemic muscle fibroses cause a tugging on the flexor muscles which produces chronic contracture.

16. **d.** The scaphoid bone is the most commonly fractured carpal bone and is at risk in a fall on the outstretched hand. The region where the physician was able to elicit tenderness describes the "anatomical snuff box." The scaphoid bone is at the base of this space. Scaphoid fractures are notoriously difficult to diagnose on X-ray exam because the injury may not be apparent until several days after the fracture is sustained. Untreated, scaphoid fractures can lead to chronic wrist pain. The styloid processes are present on the distal ends of the radius and ulna, and are not carpal bones. The pisiform and hamate are carpal bones located on the ulnar side of the hand. Injury to neither of these carpal bones would explain the location of the patient's tenderness.

CHAPTER 8: LOWER LIMB

1. **d.** Damage to the medial and lateral circumflex femoral arteries, branches of the profunda femoris, represents a serious complication of femur fractures. Although the artery to the ligamentum teres from the obturator provides some collateral circulation to this region, it is often insufficient. The femoral artery is the extension of the external iliac artery at the inguinal ligament. After giving off the profunda femoris, it courses in the anterior compartment of the thigh and has branches that provide blood supply to most of the lower limb. The internal iliac artery is a terminal branch of the common iliac. It travels posteromedial to the external iliac artery and mainly supplies structures in the pelvis.

2. **c.** The lateral femoral cutaneous nerve supplies sensory innervation to the anterolateral aspect of the upper thigh. This nerve originates from the lumbar plexus and enters the thigh deep to the inguinal ligament, where it is subject to impingement. This condition commonly results from the pressure of a protuberant abdomen on the inguinal ligament. The femoral branch of the genitofemoral nerve supplies sensory innervation to the portion of the anterior thigh over the femoral triangle. The ilioinguinal nerve supplies sensation to the anterior scrotum and medial thigh. The femoral nerve provides motor innervation to the iliacus, sartorius, pectineus, and quadriceps femoris muscles as well as cutaneous innervation to the anteromedial aspect of the thigh. All of these nerves originate from the lumbar plexus.

3. **c.** The superior gluteal nerve supplies the gluteus minimus, gluteus medius, and tensor fascia latae. These muscles abduct and medially rotate the thigh. (NOTE: It is important to recognize that, unlike the sciatic nerve, the superior gluteal nerve, together with the superior gluteal artery, exits the greater sciatic foramen *superior* to the piriformis muscle.) The gluteus maximus muscle is a powerful extensor of the thigh that receives innervation from the inferior gluteal nerve. Normal gait is relatively unaffected by injury to the gluteus maximus.

4. **a.** The maneuver described is the anterior drawer test. The test is positive, as in this case, with damage to the anterior cruciate ligament. Had you noticed unusual laxity with backward pressure to the joint, you would have suspected damage to the posterior cruciate ligament. The anterior cruciate ligament prevents hyperextension of the knee, while the posterior cruciate prevents hyperflexion. The former is the weaker of the two cruciates and hence is more commonly involved in trauma to the knee. While injury to the menisci has not been ruled out, the anterior drawer test is specifically used for the evaluation of the cruciate ligaments.

5. **a.** The chief motions of the ankle joint are dorsiflexion/plantarflexion and inversion/eversion, although some abduction/adduction

and rotation are possible. Eversion describes outward movement of the palmar aspect (sole) of the foot, whereas inversion describes an inward motion. These motions occur along the subtalar and transverse tarsal joints. Thus, in a Pott's fracture, excess eversion causes stretching of the medial deltoid ligament which initiates the subsequent chain of injuries.

6. a. The great saphenous vein courses anterior to the medial malleolus in the ankle. It then travels up the medial aspect of the leg. At the knee joint, it is approximately one handbreadth posteromedial to the epicondyle of the femur. In the thigh, the great saphenous vein dives deep to empty into the femoral vein.

7. b. In contrast to the knee joint, which involves only the tibia and the femur, both the tibia and the fibula participate in the formation of the ankle joint. The two bones articulate with the talus. The calcaneus is the largest bone of the foot that forms the heel. It does not contribute to the formation of the ankle joint.

8. c. The sartorius muscle extends from the anterior superior iliac spine to the medial aspect of the tibia. Consequently, it acts on the hip and knee joints, causing flexion of both. The name derives from the Latin word for tailor, because the action of the muscle describes the position of a tailor while performing his work. The semimembranosus and the long head of the biceps femoris are both extensors of the hip and flexors of the leg.

9. a. The adductor canal is formed by the vastus medialis, the adductor longus, and the adductor magnus. The canal provides for passage of the femoral artery and vein as well as the saphenous nerve. At the inferior aspect of the canal is the adductor hiatus, a defect in the adductor magnus muscle that provides for passage of the vessels to the posterior aspect of the knee. The femoral artery becomes the popliteal artery at this hiatus. Distal femoral artery occlusion at the adductor canal is a common cause of ischemic pain, called claudication, of the lower limb.

10. c. The dorsalis pedis is a branch of the anterior tibial artery, and hence failure to palpate a dorsalis pedis pulse might indicate occlusion of the anterior tibial artery in the leg. The dorsalis pedis may be palpated in the first interosseous space on the dorsum of the foot. The posterior tibial artery enters the foot behind the medial malleolus and divides into branches that supply the plantar surface. The peroneal (fibular) artery is a branch of the posterior tibial artery in the leg; it makes a contribution to blood supply on the dorsum of the foot. Both the anterior and posterior tibial arteries are branches of the popliteal artery. Since you are able to palpate a posterior tibial pulse, the popliteal artery is likely to be patent, and the lesion most likely is in the anterior tibial artery.

11. c. The popliteus is a muscle of the deep posterior crural compartment. It is responsible for rotating the femur laterally on the fixed tibia in order to "unlock" the knee so that motion can be initiated. The plantaris is also in the posterior crural compartment. This small muscle is a weak plantarflexor of the foot. The pectineus is a muscle in the anterior compartment of the thigh that flexes and adducts the thigh. Finally, the three peronei are muscles of the leg that evert the foot. In addition, the peroneus longus and brevis plantar flex the foot, while the peroneus tertius dorsiflexes the foot.

12. c. The infrapatellar bursa lies between the patellar ligament and the tibia. Fluid may collect in this bursa secondary to the friction caused by prolonged kneeling. The main bursa of the knee surrounds the cruciate ligaments and extends anterosuperiorly as the suprapatellar bursa. The prepatellar bursa lies between the patella and the skin. The prepatellar and infrapatellar bursae do not communicate with the joint cavity, an important point when considering the pathways for extension of infection of the joint space.

13. a. The palmar interossei in both the hand and foot are abductors. By contrast, the dorsal interossei are adductors. The interossei do not participate in flexion/extension.

14. d. The patient is unable to plantarflex his foot, suggesting impairment of the gastrocnemius, soleus, and plantaris muscles. On exam, this patient would also be expected to have a markedly diminished Achilles tendon reflex. All of these muscles are innervated by S1–S2 branches of the tibial nerve, and hence compression of the S1 nerve root is consistent with his symptoms.

15. c. The "hamstring" muscles in the posterior compartment of the thigh are the long heads of the biceps femoris, the semimembranosus, and the semitendinosus. The semimembranosus and semitendinosus tendons are prominent on the medial aspect of the knee when the joint is partially flexed. The long head of the biceps femoris inserts on the fibula. The rectus femoris, vastus medialis, vastus lateralis, and vastus intermedius comprise the anterior compartment of the thigh. Collectively, they are known as the quadriceps femoris muscles.

16. a. The femoral nerve, artery, and vein lie in close proximity in the femoral triangle. The vein is the most medial of the structures, while the nerve lies most laterally. Hence, having palpated the femoral arterial pulse, you must position the needle medially to cannulate the vein.

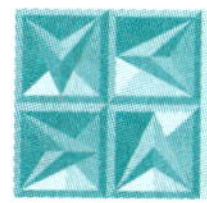

Index

UNIV DUNELM